At Sylvan, we believe that everyone can master math skills, and we are glad you have chosen our resources to help your children experience the joy of mathematics as they build crucial reasoning skills. We know that time spent reinforcing lessons learned in school will contribute to understanding and mastery.

Success in math requires more than just memorizing basic facts and algorithms; it also requires children to make connections between the real world and math concepts in order to solve problems. Successful problem solvers will be ready for the challenges of mathematics as they advance to more complex topics and encounter new problems both in school and at home.

We use a research-based, step-by-step process in teaching math at Sylvan that includes thought-provoking math problems and activities. As students increase their success as problem solvers, they become more confident. With increasing confidence, students build even more success. The design of the Sylvan workbooks lays out a roadmap for mathematical learning that is designed to lead your child to success in school.

Included with your purchase of this workbook is a coupon for a discount at a participating Sylvan center. We hope you will use this coupon to further your children's academic journeys. Let us partner with you to support the development of confident, well-prepared, independent learners.

The Sylvan Team

5th Grade Math in Action

Published in the United States by Random House, Inc., New York, and in Canada by Random House of Canada Limited, Toronto.

www.tutoring.sylvanlearning.com

Created by Smarterville Productions LLC
Producer & Editorial Direction: The Linguistic Edge
Producer: TJ Trochlil McGreevy
Writer: Amy Kraft
Cover and Interior Illustrations: Shawn Finley, Tim Goldman, and Duendes del Sur
Layout and Art Direction: SunDried Penguin
Director of Product Development: Russell Ginns

First Edition

ISBN: 978-0-375-43047-3

This book is available at special discounts for bulk purchases for sales promotions or premiums. For more information, write to Special Markets/Premium Sales, 1745 Broadway, MD 6-2, New York, New York 10019 or e-mail specialmarkets@randomhouse.com.

PRINTED IN CHINA

10 9 8 7 6 5 4 3 2 1

Contents

Contents

High Score

These four kids are all trying to get the high score in a video game where they earn points by collecting gems. MULTIPLY the gems by the correct number of points. Then ADD the scores, and CIRCLE the person with the highest score.

(dark gem) = 50 points (medium gem) = 500 points (light gem) = 5,000 points

Player 1	
800 × (dark gem)	
70 × (medium gem)	
4 × (light gem)	
Total Score	

Player 2	
1,000 × (dark gem)	
200 × (medium gem)	
1 × (light gem)	
Total Score	

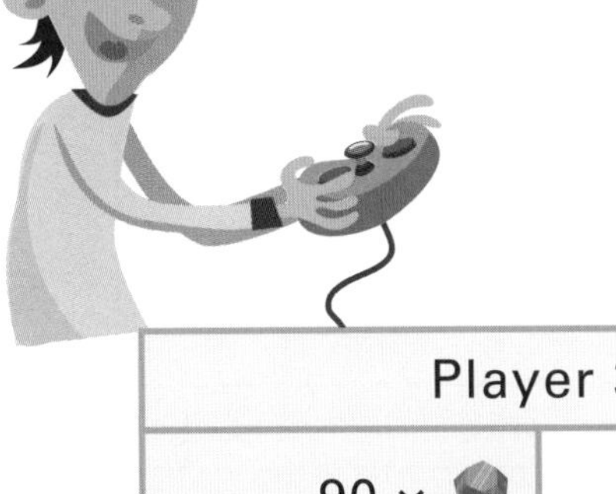

Player 3	
90 × (dark gem)	
30 × (medium gem)	
20 × (light gem)	
Total Score	

Player 4	
2,000 × (dark gem)	
11 × (medium gem)	
9 × (light gem)	
Total Score	

Spread the Word

Each person has received some important news to tell a few people. When those people hear the news, they're supposed to tell more people, and so on to spread the word. If everyone tells the news to the same number of people, WRITE the number of people who will have heard the news at each point in the chain.

1	1	1
× 3	× 4	× 5
3		
× 3	× 4	× 5
× 3	× 4	× 5
× 3	× 4	× 5

Excellent Exercise

Four friends have been working out together all week. WRITE the number of times each person did each exercise.

Maria did 25 sets of 15 jumping jacks, 18 sets of 20 sit-ups, 21 sets of 12 push-ups, and 16 sets of 10 pull-ups.

How many times did Maria do each exercise?

______ jumping jacks (1)
______ sit-ups (2)
______ push-ups (3)
______ pull-ups (4)

Al did 22 sets of 15 jumping jacks, 24 sets of 20 sit-ups, 16 sets of 12 push-ups, and 22 sets of 10 pull-ups.

How many times did Al do each exercise?

______ jumping jacks (5)
______ sit-ups (6)
______ push-ups (7)
______ pull-ups (8)

Hannah did 32 sets of 15 jumping jacks, 17 sets of 20 sit-ups, 15 sets of 12 push-ups, and 19 sets of 10 pull-ups.

How many times did Hannah do each exercise?

______ jumping jacks (9)
______ sit-ups (10)
______ push-ups (11)
______ pull-ups (12)

Jorge did 15 sets of 15 jumping jacks, 26 sets of 20 sit-ups, 26 sets of 12 push-ups, and 20 sets of 10 pull-ups.

How many times did Jorge do each exercise?

______ jumping jacks (13)
______ sit-ups (14)
______ push-ups (15)
______ pull-ups (16)

Time Travelers

Todd and Tamara are time travelers. The trouble is their time machine can only be set to travel in days and months, not weeks and years. WRITE the number of days or months they should use for each time.

1. 11 weeks = ______ days
2. 10 years = ______ months
3. 16 weeks = ______ days
4. 12 years = ______ months
5. 25 weeks = ______ days
6. 31 years = ______ months
7. 46 weeks = ______ days
8. 59 years = ______ months
9. 63 weeks = ______ days
10. 74 years = ______ months
11. 88 weeks = ______ days
12. 97 years = ______ months

Shipping Room

At the factory, action figures get packed in large boxes. There are 24 Electrogirl action figures in a box, 36 Owlboy action figures in a box, 18 Rob Roboto action figures in a box, and 45 Super Chicken action figures in a box. WRITE the number of action figures each store has ordered.

CONTAINS
18 ACTION FIGURES
ROB
ROBOTO
CONTAINS
45 ACTION FIGURES
Super
Chicken

1. Comet Comics ordered 20 boxes of Electrogirl, 35 boxes of Owlboy, 24 boxes of Rob Roboto, and 13 boxes of Super Chicken.

   ______ Electrogirl action figures

   ______ Owlboy action figures

   ______ Rob Roboto action figures

   ______ Super Chicken action figures

2. Action Figure Central ordered 17 boxes of Electrogirl, 23 boxes of Owlboy, 46 boxes of Rob Roboto, and 30 boxes of Super Chicken.

   ______ Electrogirl action figures

   ______ Owlboy action figures

   ______ Rob Roboto action figures

   ______ Super Chicken action figures

3. Lights, Camera, Action Figures! ordered 25 boxes of Electrogirl, 25 boxes of Owlboy, 25 boxes of Rob Roboto, and 25 boxes of Super Chicken.

   ______ Electrogirl action figures

   ______ Owlboy action figures

   ______ Rob Roboto action figures

   ______ Super Chicken action figures

4. Waldo's Wonderworld ordered 52 boxes of Electrogirl, 38 boxes of Owlboy, 65 boxes of Rob Roboto, and 42 boxes of Super Chicken.

   ______ Electrogirl action figures

   ______ Owlboy action figures

   ______ Rob Roboto action figures

   ______ Super Chicken action figures

Amusement Adventures

Awesome Adventure Amusement Park is collecting information about the sales of different packs of ride tickets. WRITE the total number of tickets sold.

1. 789 people each bought a 15-ticket pack. __________ tickets
2. 513 people each bought a 20-ticket pack. __________ tickets
3. 794 people each bought a 25-ticket pack. __________ tickets
4. 603 people each bought a 30-ticket pack. __________ tickets
5. 582 people each bought a 35-ticket pack. __________ tickets
6. 441 people each bought a 40-ticket pack. __________ tickets
7. 328 people each bought a 45-ticket pack. __________ tickets
8. 725 people each bought a 50-ticket pack. __________ tickets
9. 196 people each bought a 75-ticket pack. __________ tickets
10. 56 people each bought a 100-ticket pack. __________ tickets

Piles of Miles

Sky High Airways has one airplane for every pair of destinations. WRITE the total number of miles each airplane will fly in a year.

	Miles per One-Way Trip	One-Way Trips per Year	Total Miles
1. Between Columbus, OH, and Charleston, SC	557 miles	312	______
2. Between Seattle, WA, and Vancouver, BC	119 miles	730	______
3. Between Bismarck, ND, and Minneapolis, MN	470 miles	364	______
4. Between New Orleans, LA, and Houston, TX	290 miles	450	______
5. Between Boise, ID, and Portland, OR	339 miles	512	______
6. Between Denver, CO, and Calgary, AB	923 miles	328	______
7. Between Detroit, MI, and Milwaukee, WI	236 miles	674	______
8. Between Nashville, TN, and Miami FL	826 miles	242	______

That Does Not Compute!

The Great Roboto is on the fritz and is spitting out some math problems with wrong answers. CIRCLE the incorrect products.

$$\begin{array}{r} 611 \\ \times\ 153 \\ \hline 93{,}483 \end{array}$$

$$\begin{array}{r} 502 \\ \times\ 472 \\ \hline 436{,}948 \end{array}$$

$$\begin{array}{r} 819 \\ \times\ 321 \\ \hline 262{,}899 \end{array}$$

$$\begin{array}{r} 473 \\ \times\ 268 \\ \hline 126{,}854 \end{array}$$

$$\begin{array}{r} 7{,}288 \\ \times\ 52 \\ \hline 378{,}976 \end{array}$$

$$\begin{array}{r} 5{,}142 \\ \times\ 37 \\ \hline 190{,}254 \end{array}$$

$$\begin{array}{r} 8{,}994 \\ \times\ 19 \\ \hline 1{,}170{,}886 \end{array}$$

$$\begin{array}{r} 3{,}414 \\ \times\ 63 \\ \hline 715{,}072 \end{array}$$

$$\begin{array}{r} 1{,}043 \\ \times\ 916 \\ \hline 1{,}955{,}388 \end{array}$$

$$\begin{array}{r} 6{,}819 \\ \times\ 162 \\ \hline 1{,}104{,}678 \end{array}$$

$$\begin{array}{r} 2{,}446 \\ \times\ 259 \\ \hline 633{,}535 \end{array}$$

$$\begin{array}{r} 5{,}573 \\ \times\ 896 \\ \hline 4{,}993{,}408 \end{array}$$

World Tour

Sara Starlight is on a world tour performing sold-out shows in 12 different cities. ESTIMATE the number of tickets sold in each city by rounding the number of tickets to the nearest thousand.

1. In Tokyo, 4,887 tickets were sold for each of Sara Starlight's four shows. 20,000
2. In Moscow, 3,426 tickets were sold for each of Sara Starlight's three shows. ______
3. In Berlin, 6,518 tickets were sold for each of Sara Starlight's six shows. ______
4. In Copenhagen, 4,034 tickets were sold for each of Sara Starlight's four shows. ______
5. In Madrid, 3,482 tickets were sold for each of Sara Starlight's five shows. ______
6. In London, 7,245 tickets were sold for each of Sara Starlight's seven shows. ______
7. In Rome, 2,377 tickets were sold for both of Sara Starlight's two shows. ______
8. In Atlanta, 5,319 tickets were sold for each of Sara Starlight's three shows. ______
9. In New York, 6,348 tickets were sold for each of Sara Starlight's six shows. ______
10. In Chicago, 5,599 tickets were sold for each of Sara Starlight's five shows. ______
11. In Toronto, 3,692 tickets were sold for each of Sara Starlight's three shows. ______
12. In San Francisco, 7,863 tickets were sold for both of Sara Starlight's two shows. ______

Calculator Catch

Calculators can be a great help in solving problems if you push the right buttons. ESTIMATE each product by rounding to the nearest hundred, and CIRCLE the calculators showing the wrong answers.

1.

$$\begin{array}{r} 404 \\ \times\ 192 \\ \hline \end{array}$$

↓

$$\begin{array}{r} 400 \\ \times\ 200 \\ \hline 80{,}000 \end{array}$$

2.

$$\begin{array}{r} 769 \\ \times\ 329 \\ \hline \end{array}$$

↓

× ______

3.

$$\begin{array}{r} 957 \\ \times\ 521 \\ \hline \end{array}$$

↓

× ______

4.

$$\begin{array}{r} 345 \\ \times\ 466 \\ \hline \end{array}$$

↓

× ______

5.

$$\begin{array}{r} 2{,}879 \\ \times\ 289 \\ \hline \end{array}$$

↓

× ______

6.

$$\begin{array}{r} 1{,}664 \\ \times\ 743 \\ \hline \end{array}$$

↓

× ______

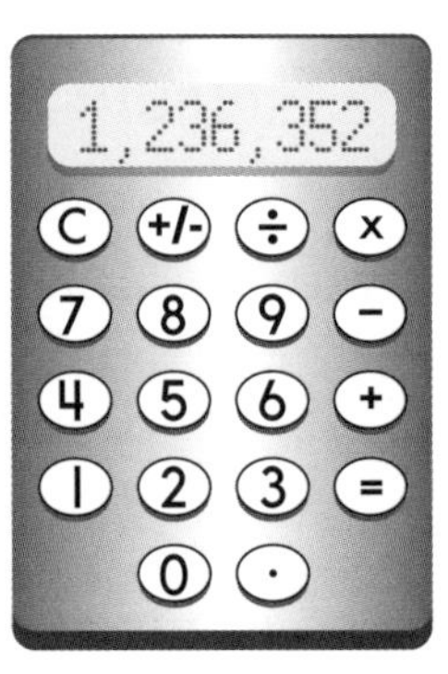

Assigned Seats

Martha has a strange way of seating guests at her parties. She gives each person a number, and each person has to find a table number that is a factor of the number. WRITE each number and name at the correct table.

Victoria: 42

Alyssa: 15

Hunter: 54

Isaac: 25

Cassandra: 45

Kenneth: 8

Faith: 66

Aaron: 32

Joseph: 10

Courtney: 16

Diana: 28

Alejandro: 18

Candy Packages

Stella's Sweet Shoppe is packing up batches of candy. Stella would like to use the largest box that can fit all of the candy with no leftovers. CIRCLE the box she should use for the batches.

HINT: Find the greatest common factor to determine how many pieces should fit in a box.

1. 64 Cinnaswirl Candies
   72 Chewy Chocolate Candies

2. 24 Nougat Nectar Candies
   42 Peanut Blast Candies

3. 35 Yummo Gummo Candies
   90 Lemon Twister Candies

4. 48 Fudgy Nut Candies
   60 Marshmallow Middles Candies

Calorie Counter

Each person eats within a range of calories every day to stay healthy. WRITE the range of calories that each person would eat in an entire year.

HINT: Remember, there are 365 days in a year.

Tammy eats between 1,460 and 1,545 calories every day.

1. In a year she'll eat between ________ and ________ calories.

Henry eats between 2,395 and 2,558 calories every day.

2. In a year he'll eat between ________ and ________ calories.

Miles eats between 1,896 and 1,970 calories every day.

3. In a year he'll eat between ________ and ________ calories.

Maya eats between 2,095 and 2,122 calories every day.

4. In a year she'll eat between ________ and ________ calories.

At the Movies

Movies on the big screen are shown at 24 frames per second, which means that for every second you're watching the movie, 24 pictures (or frames of film) move through the projector. WRITE the total number of frames for each of these movies.

HINT: First determine the number of seconds in each movie.

1. *Raiders of the Lost Park,* 94 minutes ________ frames
2. *Singing in the Train,* 114 minutes ________ frames
3. *Boy Story,* 86 minutes ________ frames
4. *Jack to the Future,* 105 minutes ________ frames
5. *Modern Mimes,* 78 minutes ________ frames
6. *Florence of Arabia,* 132 minutes ________ frames
7. *Beauty and the Feast,* 99 minutes ________ frames
8. *Paid Runner,* 121 minutes ________ frames

Buy in Bulk

Stores buy a large number of items at one time because they can usually get each item at a better price. WRITE the cost per item.

1. 10 video games for $600 $_____ per game

   100 video games for $5,000 $_____ per game

   1,000 video games for $40,000 $_____ per game

2. 6 purses for $1,800 $_____ per purse

   60 purses for $12,000 $_____ per purse

   600 purses for $60,000 $_____ per purse

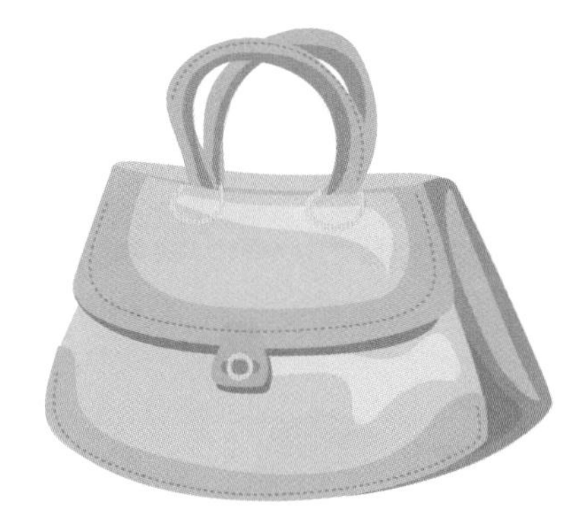

3. 10 skateboards for $2,500 $_____ per skateboard

   200 skateboards for $40,000 $_____ per skateboard

   300 skateboards for $45,000 $_____ per skateboard

4. 4 computers for $16,000 $_____ per computer

   100 computers for $350,000 $_____ per computer

   500 computers for $1,000,000 $_____ per computer

Assembly Line

Each person in the assembly line divides batches of cookies into smaller and smaller groups until the cookies are ready to be put in small packages. WRITE the number of cookies there will be in each group in the assembly line.

324	512	3,750
÷ 3	÷ 4	÷ 5
÷ 3	÷ 4	÷ 5
÷ 3	÷ 4	÷ 5
÷ 3	÷ 4	÷ 5

Cool Collections

Four kids have four cool collections, and now they're trying to organize their collections into albums. WRITE the number of album pages needed to hold each collection.

Billy has a collection of 567 baseball cards.
Each album page holds 9 baseball cards.

1. How many album pages will Billy need? ____________

Trisha has a collection of 784 photographs.
Each album page holds 4 photographs.

2. How many album pages will Trisha need? ____________

Max has a collection of 2,230 stamps.
Each album page holds 10 stamps.

3. How many album pages will Max need? ____________

Jin has a collection of 5,048 stickers.
Each album page holds 8 stickers.

4. How many album pages will Jin need? ____________

Walking Workout

Ten friends decided to see who could walk the most in a week. Everyone wore a pedometer, a device that tracks the number of steps taken. They checked in with each other at the end of seven days. WRITE the number of steps each person walked per day.

1. Elsa walked 66,045 steps. ________________ steps per day
2. Martin walked 56,896 steps. ________________ steps per day
3. Sonja walked 81,144 steps. ________________ steps per day
4. David walked 74,641 steps. ________________ steps per day
5. Brianna walked 63,056 steps. ________________ steps per day
6. Isaiah walked 55,650 steps. ________________ steps per day
7. Taylor walked 95,368 steps. ________________ steps per day
8. Garrett walked 74,956 steps. ________________ steps per day
9. Kiara walked 60,984 steps. ________________ steps per day
10. Omar walked 87,906 steps. ________________ steps per day

Family Vacation

Several families have hit the road for a driving vacation. If each family didn't vary its speed for the whole trip, WRITE the number of hours each family drove.

1. Peter's family drove 420 miles going 60 miles per hour. _____ hours
2. Sandy's family drove 210 miles going 35 miles per hour. _____ hours
3. Dane's family drove 405 miles going 45 miles per hour. _____ hours
4. Ella's family drove 550 miles going 50 miles per hour. _____ hours
5. Ricardo's family drove 624 miles going 48 miles per hour. _____ hours
6. Maya's family drove 456 miles going 57 miles per hour. _____ hours
7. Chen's family drove 160 miles going 32 miles per hour. _____ hours
8. Justine's family drove 522 miles going 29 miles per hour. _____ hours
9. Hurley's family drove 236 miles going 59 miles per hour. _____ hours
10. Claire's family drove 378 miles going 42 miles per hour. _____ hours
11. Austin's family drove 636 miles going 53 miles per hour. _____ hours
12. Karen's family drove 616 going 44 miles per hour. _____ hours

Avid Readers

WRITE the number of days it will take these friends to read each of their books, and the number of pages that will be left to read on the last day.

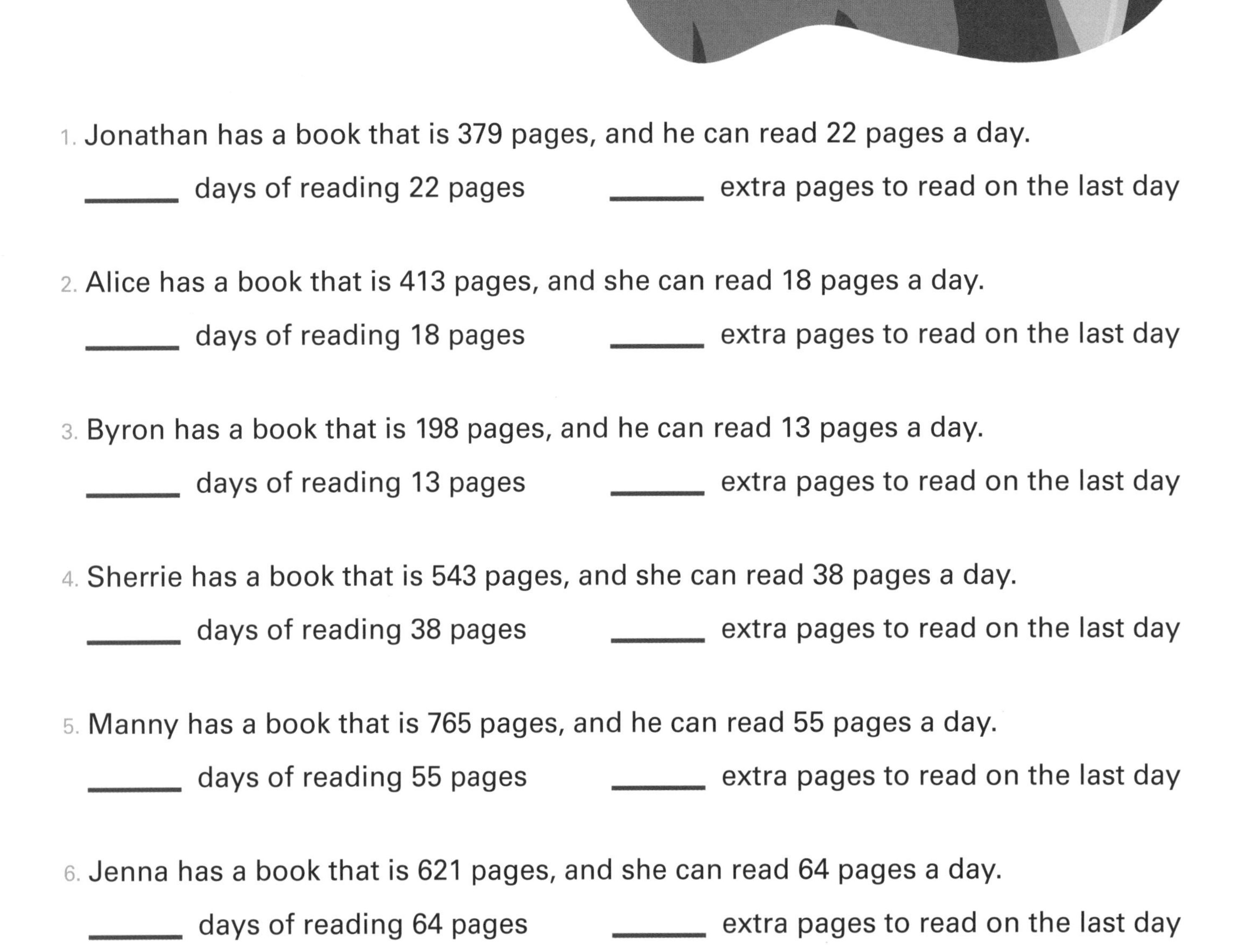

1. Jonathan has a book that is 379 pages, and he can read 22 pages a day.

_____ days of reading 22 pages _____ extra pages to read on the last day

2. Alice has a book that is 413 pages, and she can read 18 pages a day.

_____ days of reading 18 pages _____ extra pages to read on the last day

3. Byron has a book that is 198 pages, and he can read 13 pages a day.

_____ days of reading 13 pages _____ extra pages to read on the last day

4. Sherrie has a book that is 543 pages, and she can read 38 pages a day.

_____ days of reading 38 pages _____ extra pages to read on the last day

5. Manny has a book that is 765 pages, and he can read 55 pages a day.

_____ days of reading 55 pages _____ extra pages to read on the last day

6. Jenna has a book that is 621 pages, and she can read 64 pages a day.

_____ days of reading 64 pages _____ extra pages to read on the last day

Shipping Room

At the factory, food gets packed in large boxes. WRITE the number of boxes needed for each type of food and the amount of food that's left over.

1. The factory made 568 jars of peanut butter, which come in boxes of 12 jars.

   ______ boxes ______ jars left over

2. The factory made 912 juice boxes, which come in boxes of 32 juice boxes.

   ______ boxes ______ juice boxes left over

3. The factory made 745 packages of spaghetti, which come in boxes of 20 packages.

   ______ boxes ______ packages left over

4. The factory made 838 cans of soup, which come in boxes of 18 cans.

   ______ boxes ______ cans left over

5. The factory made 2,405 frozen pizzas, which come in boxes of 15 pizzas.

   ______ boxes ______ pizzas left over

6. The factory made 5,971 bags of rice, which come in boxes of 12 bags.

   ______ boxes ______ bags left over

That Does Not Compute!

The Great Roboto is on the fritz and is spitting out some math problems with wrong answers. CIRCLE the incorrect quotients.

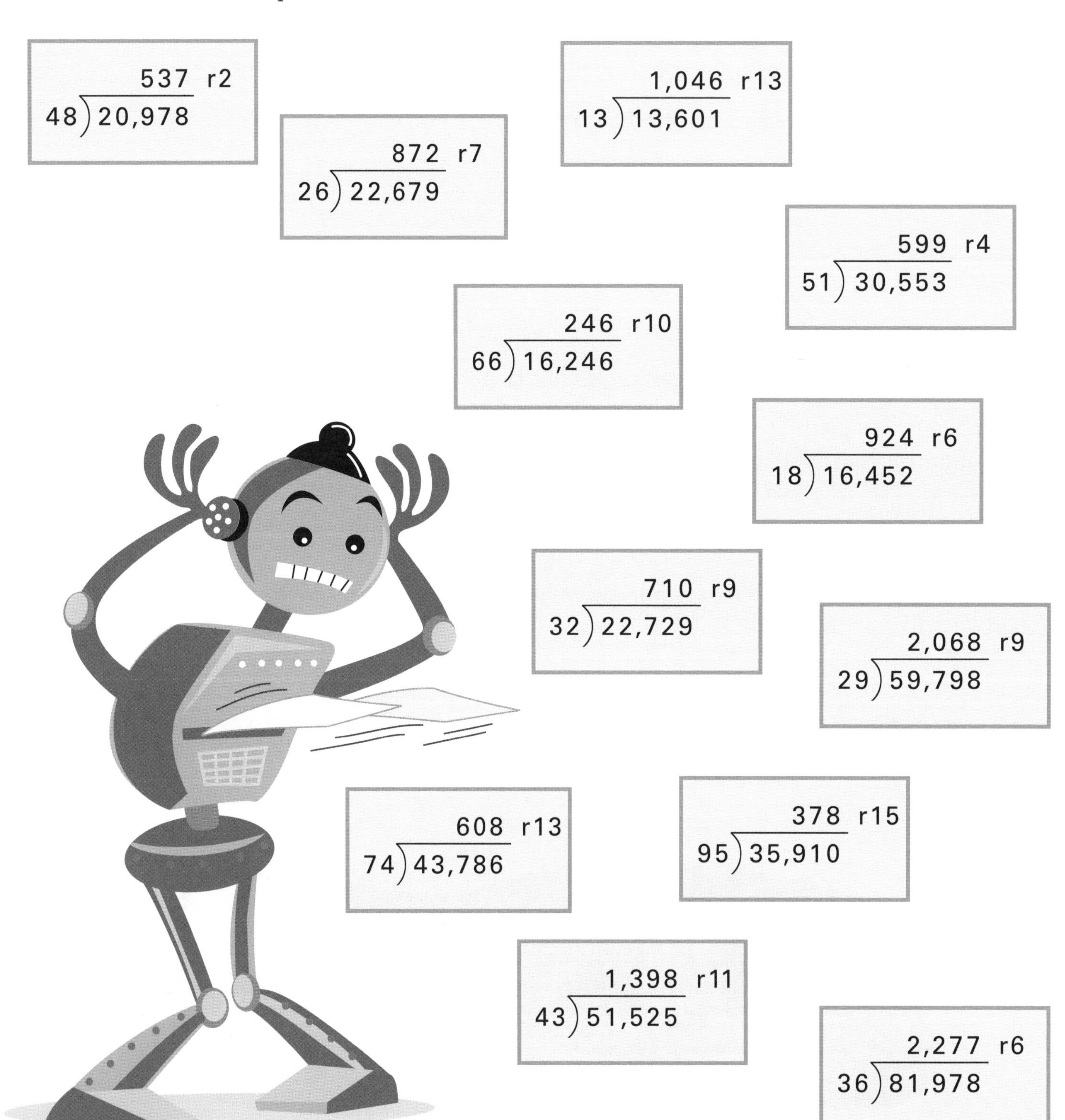

Car Share

Different groups of people are talking about buying a car that they could all share. ESTIMATE the cost per person by rounding the cost of the car to the nearest thousand.

1. Six people are talking about splitting the cost of a compact car priced at $18,376. $ ____________ per person

2. Four people are thinking about splitting the cost of a used car priced at $7,823. $ ____________ per person

3. Eight people are talking about buying a $71,671 SUV. $ ____________ per person

4. Five people are talking about buying a hybrid car for $25,469. $ ____________ per person

5. Seven people are thinking about buying a $27,544 minivan. $ ____________ per person

6. Three people are going to buy a convertible sports car for $104,762. $ ____________ per person

Calculator Catch

Calculators can be a great help in solving problems if you push the right buttons. ESTIMATE each quotient by rounding the dividend to the nearest thousand, and CIRCLE the calculators showing the wrong answers.

1. 9)36,423

2. 15)44,835

3. 3)65,763

4. 4)20,328

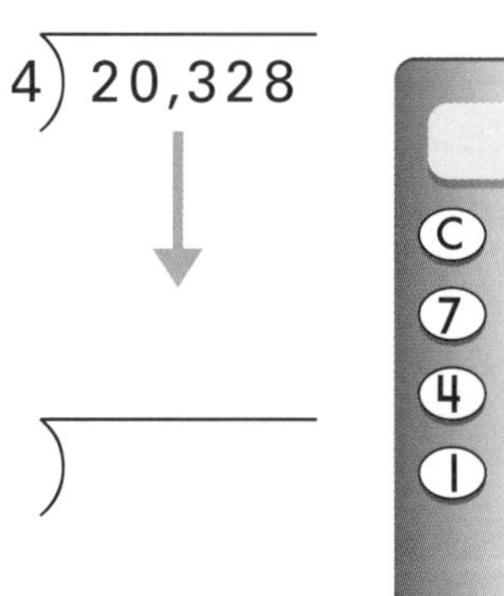

5. 24)48,144

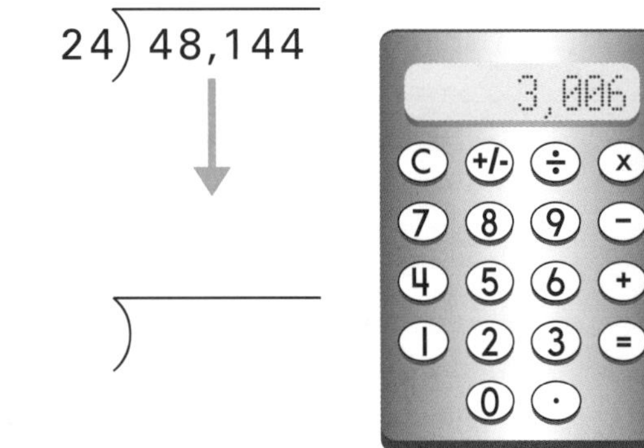

6. 70)69,580

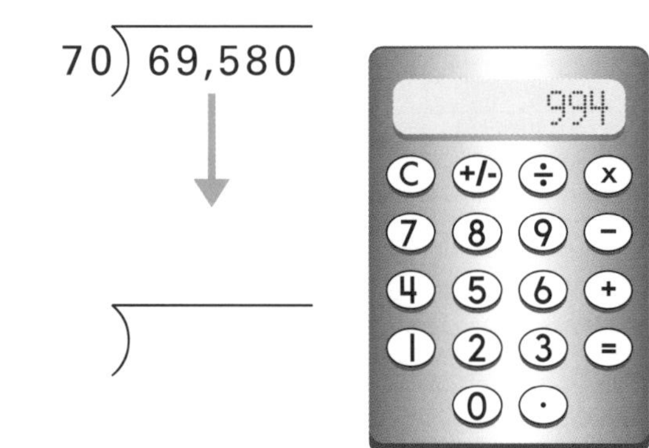

Factor Families

A **prime number** can only be divided evenly by itself and 1. A **composite number** has more factors than itself and 1. WRITE the missing numbers in each factor family.

Example:

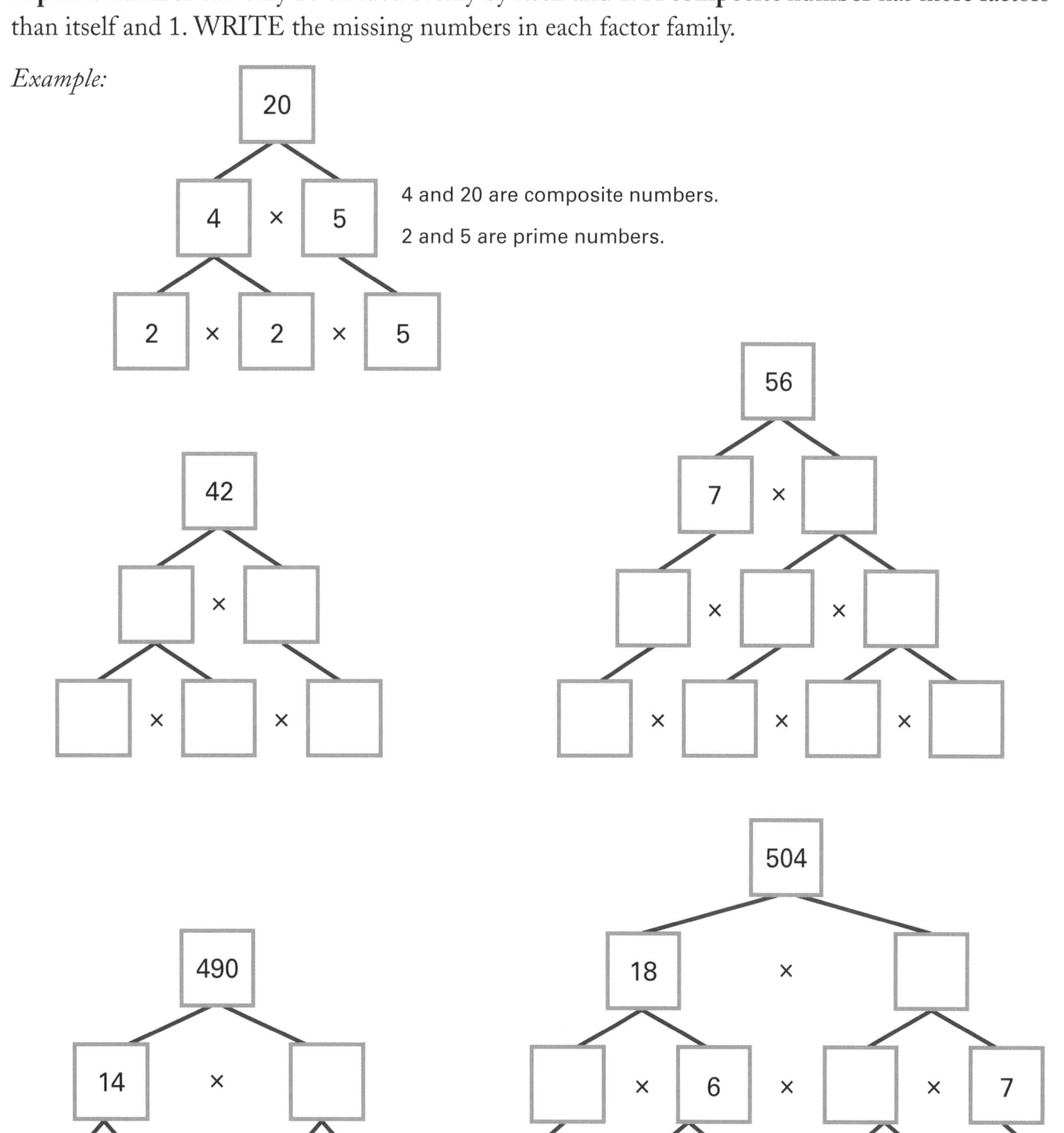

Prime Find

Mikey is trying to find all of the prime numbers between 1 and 100. CIRCLE each prime number.

1	2	3	4	5	6	7	8	9	10
11	12	13	14	15	16	17	18	19	20
21	22	23	24	25	26	27	28	29	30
31	32	33	34	35	36	37	38	39	40
41	42	43	44	45	46	47	48	49	50
51	52	53	54	55	56	57	58	59	60
61	62	63	64	65	66	67	68	69	70
71	72	73	74	75	76	77	78	79	80
81	82	83	84	85	86	87	88	89	90
91	92	93	94	95	96	97	98	99	100

Cool Collections

Each album holds 100 pages. WRITE the number of album pages needed for each collection. Then WRITE the number of full albums for the collection, and the number of full pages in the last album.

1. Brenda has a collection of 4,104 photographs. Each album page holds 8 photographs.

How many album pages will Brenda need? ________

How many full albums will Brenda have? ________

How many extra album pages will Brenda have? ________

2. Sam has a collection of 11,502 baseball cards. Each album page holds 18 baseball cards.

How many album pages will Sam need? ________

How many full albums will Sam have? ________

How many extra album pages will Sam have? ________

3. Maya has a collection of 19,824 stickers. Each album page holds 24 stickers.

How many album pages will Maya need? ________

How many full albums will Maya have? ________

How many extra album pages will Maya have? ________

4. Wyatt has a collection of 32,688 stamps. Each album page holds 36 stamps.

How many album pages will Wyatt need? ________

How many full albums will Wyatt have? ________

How many extra album pages will Wyatt have? ________

Remember the Remainders

READ each paragraph, and WRITE the answers.

Eight families chipped in to throw a neighborhood block party that included a big barbecue, a bounce house, and a live band. The cost of the party was $7,451, and each family has paid $904.

1. How much of the party cost has not been paid? $________

Emily has been making bags of 36 jellybeans to give away along the parade route. She started with 18,025 jellybeans and made 500 bags.

2. How many jellybeans did not get put into bags? ________

The state government has purchased 64,123 new books for the state's 53 libraries. Each library received 1,209 books.

3. How many books remain? ________

Walter won $35,782 on a game show. He has decided that he wants to give an equal amount to each of his 12 favorite charities, and after that he'd keep whatever is left over. Walter gave $2,981 to each charity.

4. How much will he keep? $________

Best Price

CIRCLE the item with the lowest price.

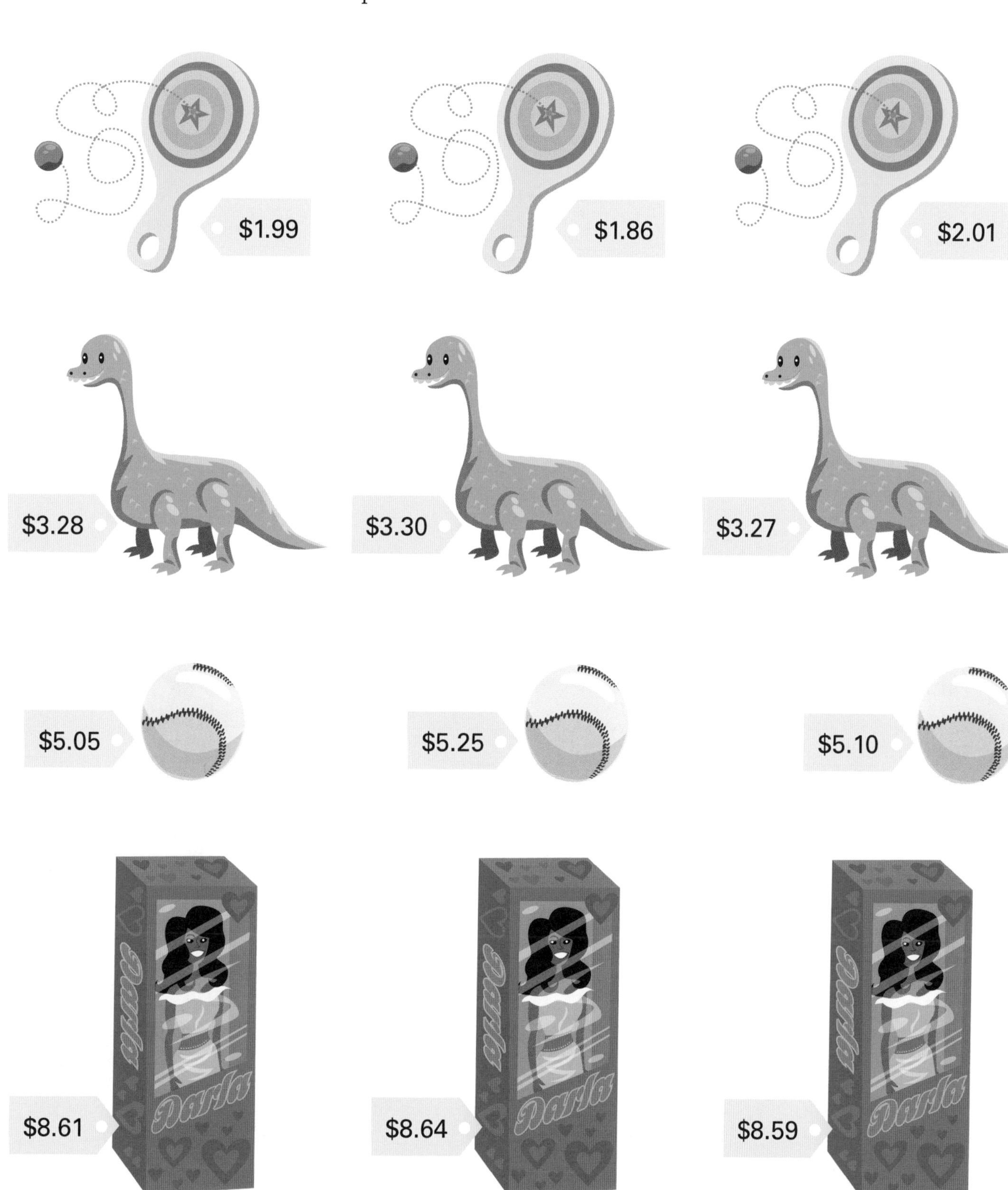

Batter Up

In baseball, the batting average is a number that shows the number of hits a batter got versus the number of times the batter was at bat. The higher the number, the better the batting average. WRITE the numbers 1 through 10 so that 1 is the best batting average and 10 is the worst batting average.

Player	Batting Average	Rank
1. Sly Young	.251	_____
2. Larry Bonds	.278	_____
3. Josephine Jackson	.349	_____
4. Willie Frays	.318	_____
5. Melanie Ott	.185	_____
6. Ruth Babe	.272	_____
7. Ned Williams	.367	_____
8. Jill Hamilton	.299	_____
9. Frank Aaron	.205	_____
10. Peg Maddux	.218	_____

Piggy Bank

ROUND the amount of money in each piggy bank to the nearest dollar.

HINT: When rounding, look at the digit to the right of the place you're rounding to. If that digit is less than 5, round down. If it is 5 or greater, round up.

$________
1

$________
2

$________
3

$________
4

$________
5

$________
6

$________
7

$________
8

Track and Field

Athletes have come from all over the country to participate in the country's track-and-field events, including shot put, long jump, and the 100-meter hurdles.

ROUND each shot-put score to the nearest meter.

1. 17.43 m ______ m
2. 22.81 m ______ m
3. 21.33 m ______ m
4. 18.62 m ______ m
5. 20.49 m ______ m
6. 21.55 m ______ m

ROUND each long-jump score to the nearest tenth of a meter.

7. 7.21 m ______ m
8. 5.87 m ______ m
9. 6.55 m ______ m
10. 5.98 m ______ m
11. 6.74 m ______ m
12. 7.42 m ______ m

ROUND each hurdle score to the nearest hundredth of a second.

13. 12.932 sec ________ sec
14. 13.048 sec ________ sec
15. 12.876 sec ________ sec
16. 14.215 sec ________ sec
17. 13.494 sec ________ sec
18. 12.723 sec ________ sec

Plus Tax

WRITE the total cost of each item when sales tax is added.

1.

Price	$84.50
Tax	$ 5.49
Total	$

2.

Price	$5.47
Tax	$0.36
Total	$

3.

Price	$159.97
Tax	$ 10.40
Total	$

4.

Price	$575.65
Tax	$ 37.42
Total	$

5.

Price	$27.68
Tax	$ 1.80
Total	$

6.

Price	$229.99
Tax	$ 14.95
Total	$

Ask the Judges

The winner of the dance competition is the one with the highest score out of 10 possible points. ADD each dancer's scores. Then CIRCLE the winning dancer.

Best Price

Janey is shopping around for the best price on the new must-have pair of Stargazer Sneakers. She's found them for different prices at four different stores, and Janey was able to find a coupon for each store. CIRCLE the sneakers with the best price if Janey uses the coupon.

That Does Not Compute!

The Great Roboto is on the fritz and is spitting out some math problems with wrong answers. CIRCLE the incorrect differences.

$$\begin{array}{r} 7.04 \\ -\ 3.9 \\ \hline 3.14 \end{array}$$

$$\begin{array}{r} 78.131 \\ -\ 4.226 \\ \hline 73.905 \end{array}$$

$$\begin{array}{r} 913.568 \\ -834.92 \\ \hline 78.648 \end{array}$$

$$\begin{array}{r} 15.26 \\ -\ 9.87 \\ \hline 6.39 \end{array}$$

$$\begin{array}{r} 20.8 \\ -16.16 \\ \hline 4.62 \end{array}$$

$$\begin{array}{r} 201.55 \\ -166.65 \\ \hline 34.95 \end{array}$$

$$\begin{array}{r} 0.12 \\ -0.108 \\ \hline 0.012 \end{array}$$

$$\begin{array}{r} 99.9 \\ -88.88 \\ \hline 11.11 \end{array}$$

$$\begin{array}{r} 425.2 \\ -\ 58.29 \\ \hline 366.91 \end{array}$$

$$\begin{array}{r} 300.03 \\ -\ 30.003 \\ \hline 270.27 \end{array}$$

$$\begin{array}{r} 27.51 \\ -15.72 \\ \hline 12.79 \end{array}$$

$$\begin{array}{r} 9.6 \\ -2.007 \\ \hline 7.593 \end{array}$$

Copy Shop

Coco's Copy Shop has different prices for copies depending on how many copies you make. The chart shows the price per copy. WRITE the cost of each copy order.

Number of Copies	Black-and-White Copies	Color Copies
0–25	$0.10	$0.28
26–50	$0.08	$0.25
51–100	$0.06	$0.21
101–500	$0.05	$0.18
501–1,000	$0.04	$0.12

1. Suzie needs to make 20 color copies of her garage-sale sign.

   $________

2. Brad needs to make 6 black-and-white copies of his latest short story.

   $________

3. Alison needs to make 80 black-and-white copies of the program for the school play.

   $________

4. Brent wants to make 33 color copies of a vacation photograph he took.

   $________

5. Margaret needs to make 175 black-and-white copies of a flyer to sell her bike.

   $________

6. Kevin wants to make copies of the comic strip he drew. He'll need a total of 375 color copies.

   $________

7. Adina wants to make 642 black-and-white copies of a poem she wrote.

   $________

8. Dan needs to make 558 color copies of the flyer for his band's show next week.

   $________

Everyday Electricity

The electric company charges $0.092 for every kilowatt hour (kWh) of use. One kilowatt hour lights a 100-watt light bulb for about 10 hours. WRITE the cost of the electrical use for each month, rounding to the nearest penny where necessary.

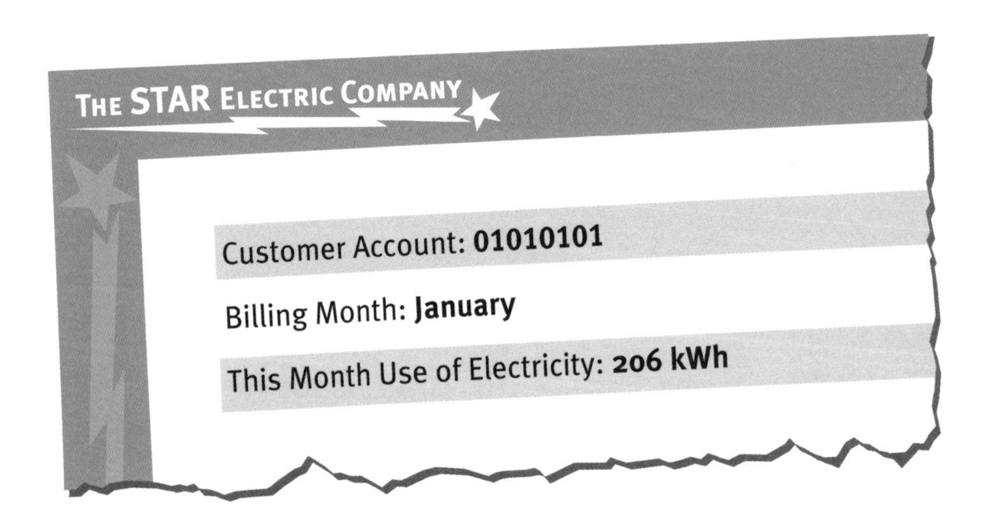

1. January: 206 kWh $________
2. February: 275 kWh $________
3. March: 284 kWh $________
4. April: 320 kWh $________
5. May: 349 kWh $________
6. June: 513 kWh $________
7. July: 684 kWh $________
8. August: 736 kWh $________
9. September: 415 kWh $________
10. October: 302 kWh $________
11. November: 266 kWh $________
12. December: 199 kWh $________

Snack Bar

Groups of friends are chipping in to buy snacks at the snack bar. WRITE the cost for each person.

1. Four friends are chipping in for nachos that cost $5.48.

   Each person will pay $__________

2. Two friends are splitting a milkshake and a bag of pretzels for $6.72.

   Each person will pay $__________

3. Six friends are chipping in for three hot dogs and three lemonades for a total of $11.04.

   Each person will pay $__________

4. Five friends are splitting two orders of cheeseburgers with fries that have a total cost of $12.45.

   Each person will pay $__________

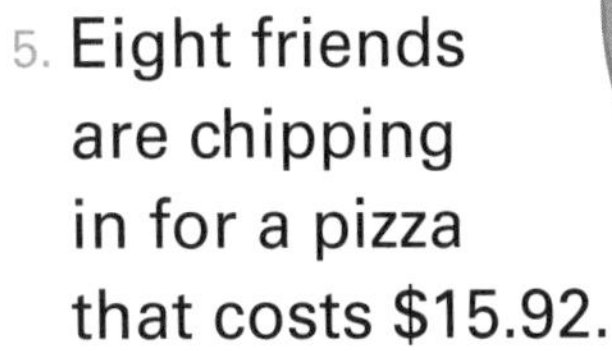

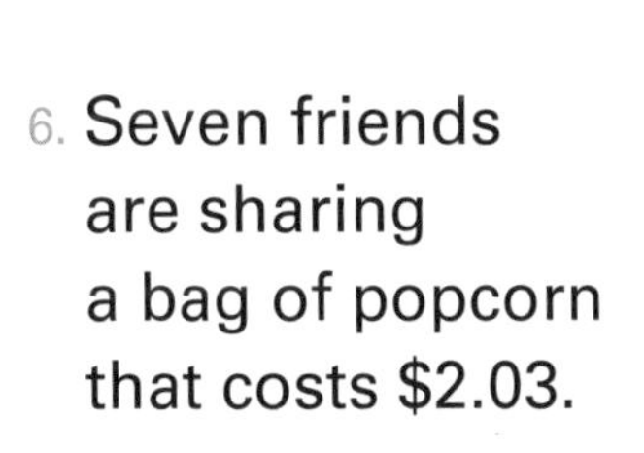

5. Eight friends are chipping in for a pizza that costs $15.92.

   Each person will pay $__________

6. Seven friends are sharing a bag of popcorn that costs $2.03.

   Each person will pay $__________

That Does Not Compute!

The Great Roboto is on the fritz and is spitting out some math problems with wrong answers. CIRCLE the incorrect quotients.

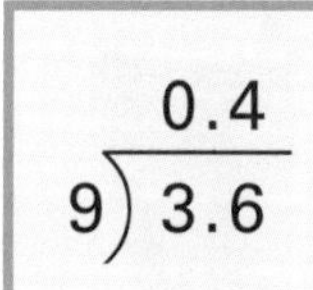

$\begin{array}{r} 0.4 \\ 9\overline{)3.6} \end{array}$

$\begin{array}{r} 0.6 \\ 8\overline{)5.6} \end{array}$

$\begin{array}{r} 0.14 \\ 12\overline{)1.44} \end{array}$

$\begin{array}{r} 1.08 \\ 4\overline{)4.32} \end{array}$

$\begin{array}{r} 1.39 \\ 7\overline{)97.3} \end{array}$

$\begin{array}{r} 2.5 \\ 5\overline{)12.5} \end{array}$

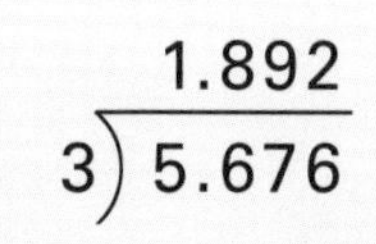

$\begin{array}{r} 1.892 \\ 3\overline{)5.676} \end{array}$

$\begin{array}{r} 4.5 \\ 10\overline{)0.45} \end{array}$

$\begin{array}{r} 8.23 \\ 15\overline{)123.45} \end{array}$

$\begin{array}{r} 0.28 \\ 23\overline{)0.644} \end{array}$

$\begin{array}{r} 1.49 \\ 17\overline{)23.953} \end{array}$

$\begin{array}{r} 3.68 \\ 40\overline{)147.2} \end{array}$

Pay the Check

WRITE the total of each restaurant check before and after the tax. Then, WRITE the change each person would get back if paying with a 100-dollar bill.

1.

GUEST CHECK

Date	Table	Guests	Server	000901

Shrimp scampi	$18.20
Rigatoni with broccoli	$11.99
Spaghetti and meatballs	$15.50
Food Total	
Tax	$ 8.22
Total	

Change: $____.____

2.

GUEST CHECK

Date	Table	Guests	Server	000902

Prime rib	$22.95
Roast chicken	$17.95
Sirloin steak	$29.95
Food Total	
Tax	$14.17
Total	

Change: $____.____

3.

GUEST CHECK

Date	Table	Guests	Server	000903

Curry stew	$13.80
Spicy rice noodles	$18.25
Spinach wonton soup	$ 6.88
Food Total	
Tax	$ 5.84
Total	

Change: $____.____

4.

GUEST CHECK

Date	Table	Guests	Server	000904

Tomato and feta salad	$10.35
Lamb shish kebab	$18.95
Chicken stew	$15.40
Club sandwich	$13.24
Food Total	
Tax	$11.59
Total	

Change: $____.____

Calculator Catch

Calculators can be a great help in solving problems if you push the right buttons. WRITE each answer, then CIRCLE the calculators showing the wrong answers.

1. $3.926 + 0.187$

2. $216.8 + 35.91$

3. $72.516 - 9.828$

4. $14.5 - 5.93$

5. 87.03×9

6. 15.84×1.7

7. $18\overline{)0.126}$

8. $21\overline{)561.12}$

Delicious Duos

Equivalent fractions are fractions that have the same value. CIRCLE the two foods in each row that have the equivalent fraction of food.

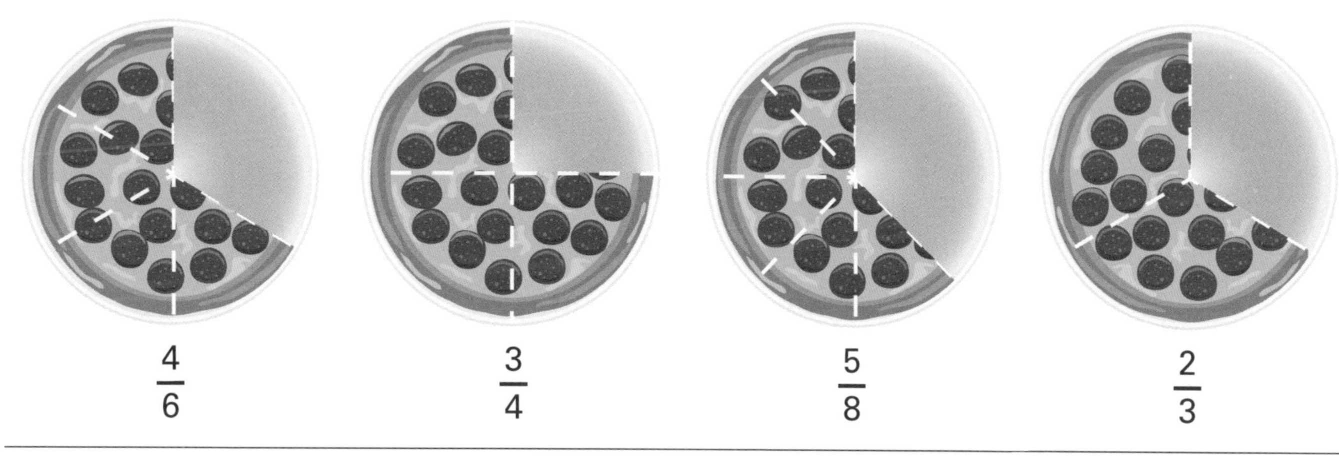

$\frac{9}{12}$ $\frac{5}{8}$ $\frac{3}{4}$ $\frac{7}{9}$

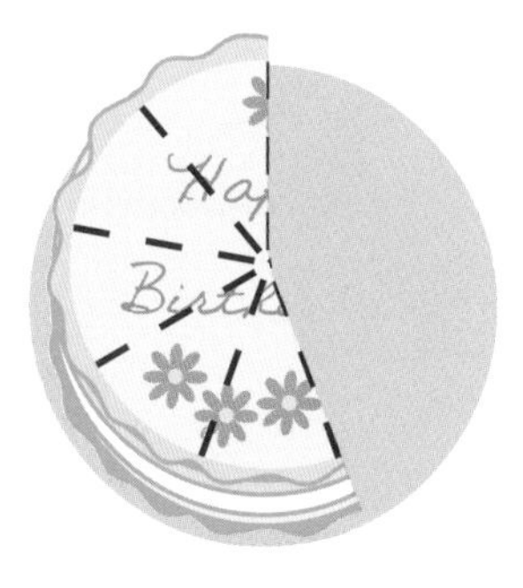

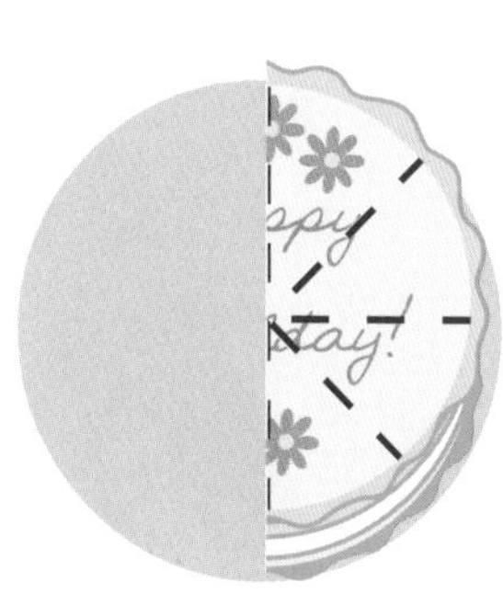

$\frac{5}{9}$ $\frac{1}{2}$ $\frac{5}{6}$ $\frac{4}{8}$

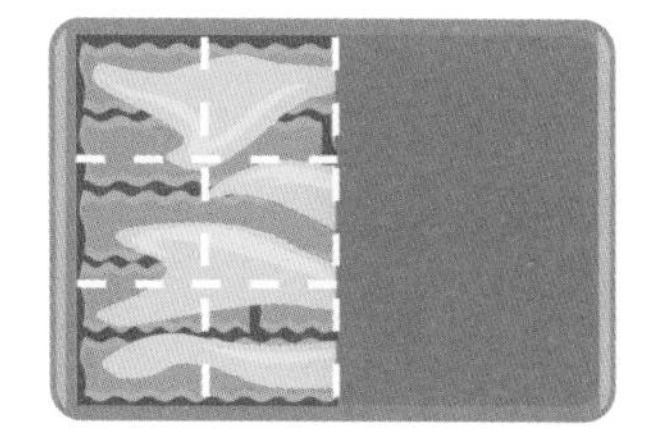

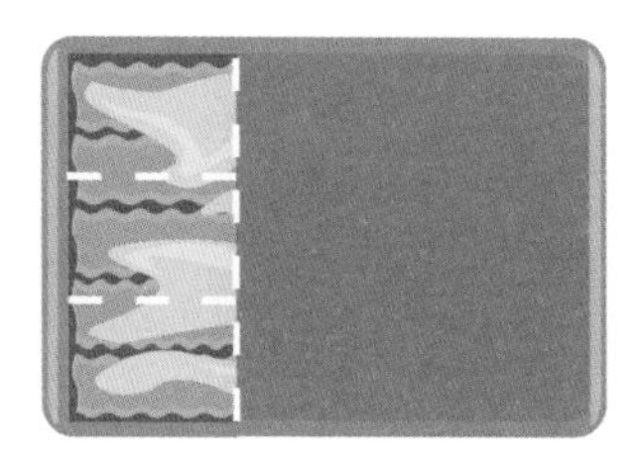

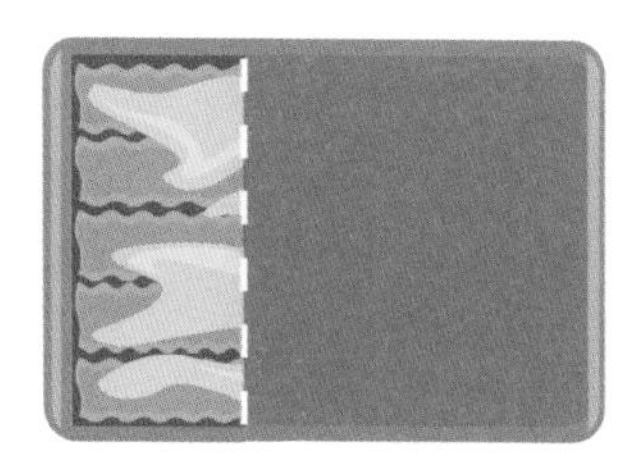

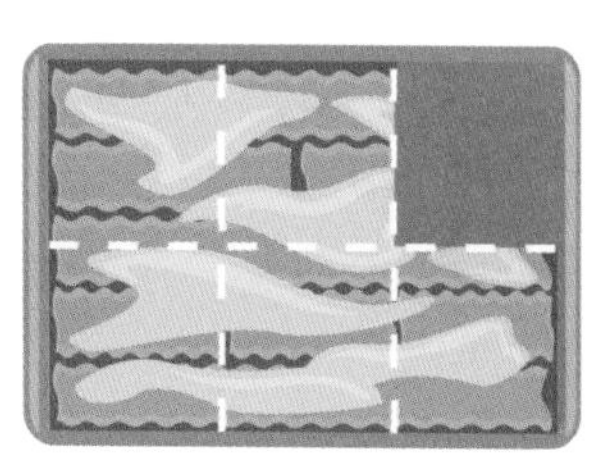

$\frac{6}{12}$ $\frac{3}{9}$ $\frac{1}{3}$ $\frac{5}{6}$

Kate's Kitchen

Today in Kate's Kitchen, Kate is showing how to make measurements when you don't have the right size measuring cup. WRITE the equivalent measurement.

HINT: Fractions that have the same number in the numerator and the denominator are equivalent to 1.

1. 1 cup = $\frac{2}{2}$ cup
2. 1 cup = $\frac{}{8}$ cup
3. 1 cup = $\frac{}{4}$ cup
4. $\frac{3}{4}$ cup = $\frac{}{8}$ cup
5. 1 cup = $\frac{}{3}$ cup
6. $\frac{1}{2}$ cup = $\frac{}{4}$ cup
7. $\frac{2}{3}$ cup = $\frac{}{6}$ cup
8. $\frac{1}{4}$ cup = $\frac{}{8}$ cup

At the Fair

A fraction is in its simplest form if the only common factor of the numerator and the denominator is 1. WRITE each fraction in the paragraph as a fraction in its simplest form.

Ethan and his family head to the county fair, one of Ethan's favorite places. When they arrive, $\frac{6}{8}$ of the parking lots are already full, so they park in lot 7. Ethan's parents buy 12 game tickets for the kids, and give $\frac{8}{12}$ of the tickets to Ethan and $\frac{4}{12}$ of the tickets to his little sister, Nora. Ethan uses $\frac{6}{8}$ of his tickets to play the ring toss. He uses the other $\frac{2}{8}$ of his tickets trying to knock down bottles with a baseball. The most he knocks down is $\frac{3}{9}$ of the bottles. Ethan and Nora split a bag of six pretzel twists. They each take $\frac{3}{6}$ of the bag. Before it's time to go home, Ethan and his family sit down to enjoy a show with 10 different acts. Ethan likes the $\frac{4}{10}$ of the show in which people tell jokes, but Nora likes the $\frac{6}{10}$ of the show that features musical acts. After a great day at the fair, it's time to go home.

1. $\frac{3}{4}$
2. $\frac{\quad}{\quad}$
3. $\frac{\quad}{\quad}$
4. $\frac{\quad}{\quad}$
5. $\frac{\quad}{\quad}$
6. $\frac{\quad}{\quad}$
7. $\frac{\quad}{\quad}$
8. $\frac{\quad}{\quad}$

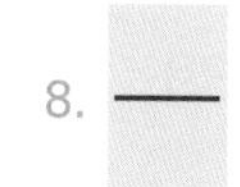

9. $\frac{\quad}{\quad}$

Amusement Adventures

The people at Rocket Launch Amusement Park want to make some changes, so they asked visitors what they thought about the amusement park. WRITE each fraction in its simplest form.

1. What is your favorite ride?
   Survey of 24 people

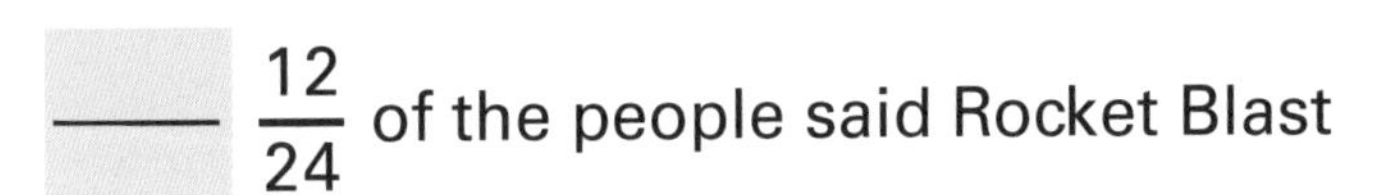
____ $\frac{12}{24}$ of the people said Rocket Blast

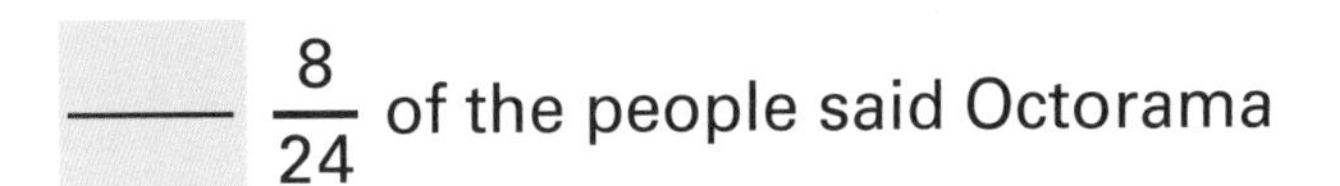
____ $\frac{8}{24}$ of the people said Octorama

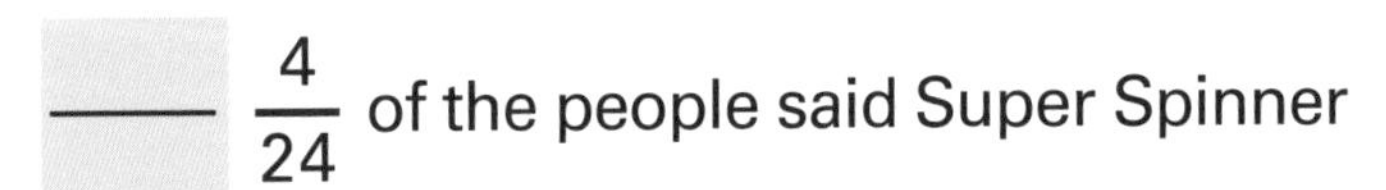
____ $\frac{4}{24}$ of the people said Super Spinner

2. What is your favorite food?
   Survey of 18 people

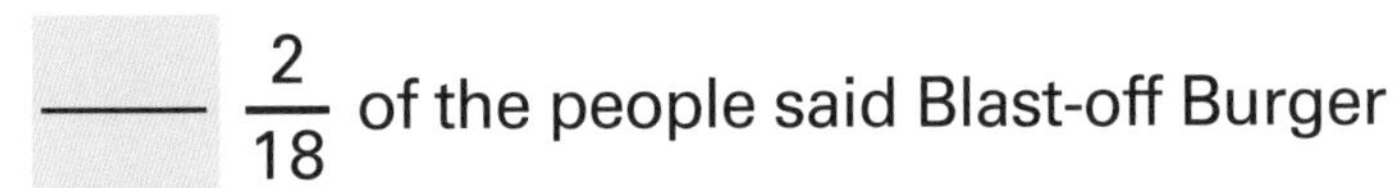
____ $\frac{2}{18}$ of the people said Blast-off Burger

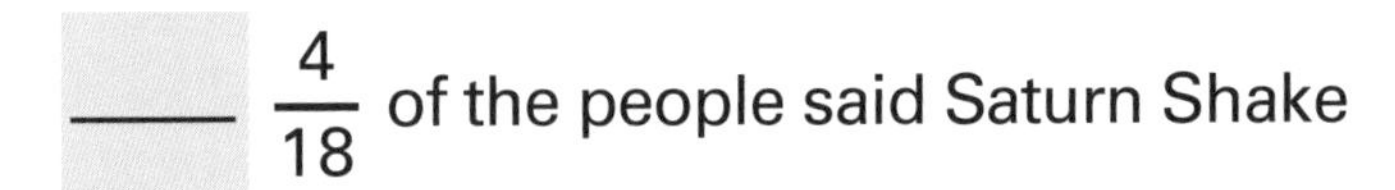
____ $\frac{4}{18}$ of the people said Saturn Shake

____ $\frac{12}{18}$ of the people said Launch Fries

3. How many tickets did you buy?
   Survey of 30 people

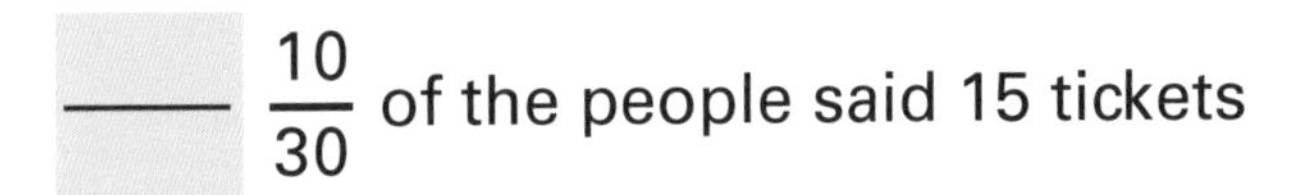
____ $\frac{10}{30}$ of the people said 15 tickets

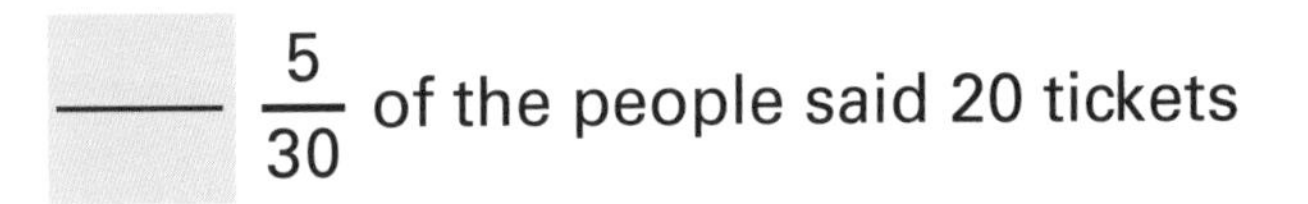
____ $\frac{5}{30}$ of the people said 20 tickets

____ $\frac{15}{30}$ of the people said 25 tickets

4. What is the longest you waited for a ride?
   Survey of 36 people

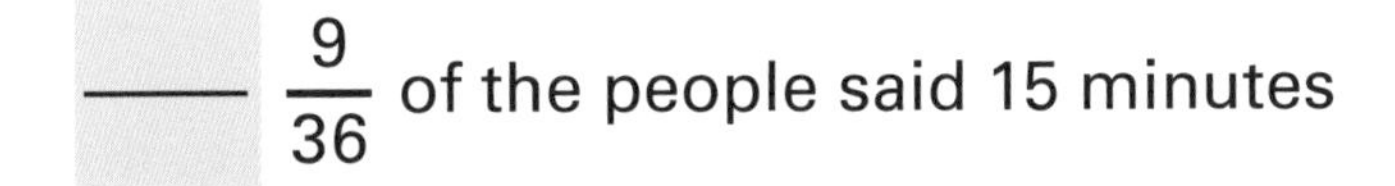
____ $\frac{9}{36}$ of the people said 15 minutes

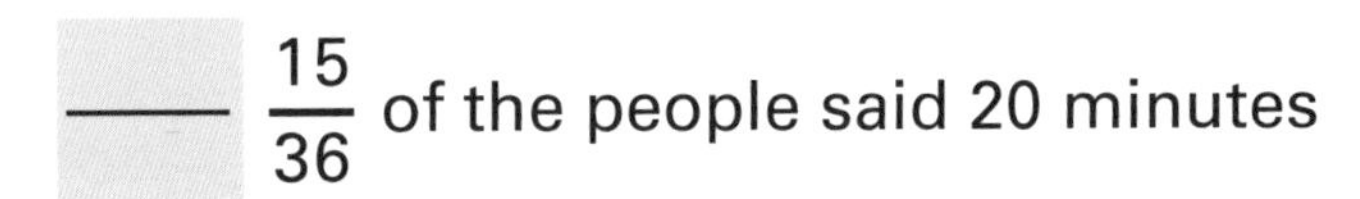
____ $\frac{15}{36}$ of the people said 20 minutes

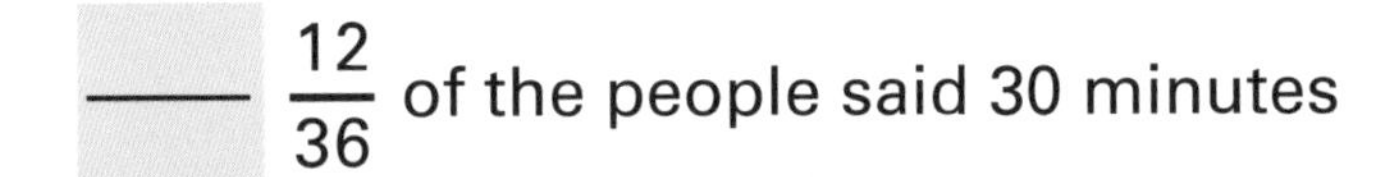
____ $\frac{12}{36}$ of the people said 30 minutes

Rainy Days

A rain gauge measures the amount of rain that falls. WRITE each measurement as a mixed number.

1. April 1 $\frac{4}{3}$ in. ________ in.
2. April 6 $\frac{8}{5}$ in. ________ in.
3. April 7 $\frac{5}{2}$ in. ________ in.
4. April 10 $\frac{7}{6}$ in. ________ in.
5. April 13 $\frac{9}{4}$ in. ________ in.
6. April 18 $\frac{19}{16}$ in. ________ in.
7. April 22 $\frac{15}{8}$ in. ________ in.
8. April 29 $\frac{23}{10}$ in. ________ in.

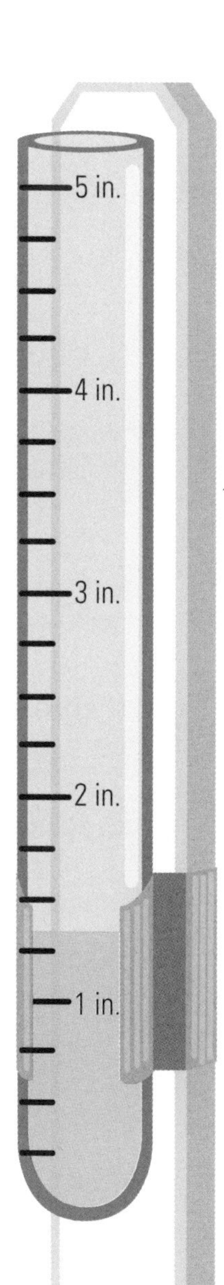

Gassing Up

How much gasoline did each person get at the gas station? WRITE the amount as a mixed number.

1. Kate put $\frac{25}{4}$ gallons into her car. ________ gallons
2. Alberto put $\frac{27}{4}$ gallons into his car. ________ gallons
3. Juliet put $\frac{31}{3}$ gallons into her car. ________ gallons
4. Benjamin put $\frac{53}{6}$ gallons into his car. ________ gallons
5. Sheila put $\frac{15}{2}$ gallons into her car. ________ gallons
6. Brian put $\frac{35}{3}$ gallons into his car. ________ gallons
7. Jackie put $\frac{17}{4}$ gallons into her car. ________ gallons
8. Russell put $\frac{83}{8}$ gallons into his car. ________ gallons

Kate's Kitchen

Today in Kate's Kitchen, Kate is showing how to make measurements when you don't have the right kind of measuring cup. WRITE the equivalent measurement as an improper fraction.

1. $1\frac{1}{2}$ cups = $\frac{3}{2}$ cups
2. $3\frac{1}{3}$ cups = $\frac{\quad}{3}$ cups
3. $2\frac{1}{3}$ cups = $\frac{\quad}{3}$ cups
4. $5\frac{1}{2}$ cups = $\frac{\quad}{2}$ cups
5. $1\frac{3}{4}$ cups = $\frac{\quad}{4}$ cups
6. $2\frac{1}{8}$ cups = $\frac{\quad}{8}$ cups
7. $4\frac{2}{3}$ cups = $\frac{\quad}{3}$ cups
8. $1\frac{7}{8}$ cups = $\frac{\quad}{8}$ cups

1

1/2

Leftover Pizza

Each amount of leftover pizza is shown as a mixed number. WRITE the amount of leftover pizza as an improper fraction.

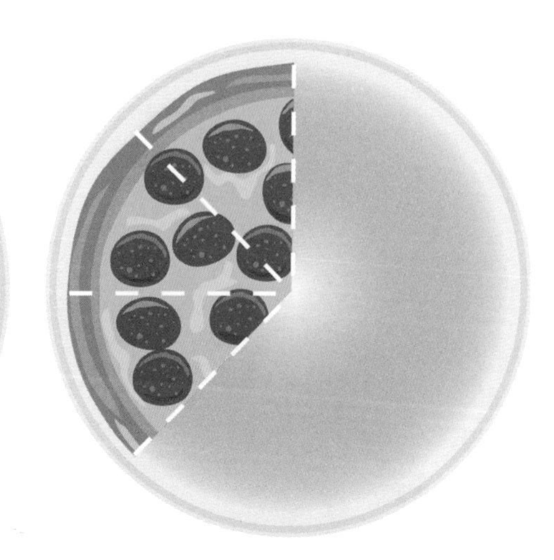

1. $2\frac{3}{8}$ pizzas = ________ pizzas
2. $1\frac{5}{6}$ pizzas = ________ pizzas
3. $3\frac{3}{4}$ pizzas = ________ pizzas
4. $1\frac{9}{10}$ pizzas = ________ pizzas
5. $3\frac{1}{3}$ pizzas = ________ pizzas
6. $2\frac{1}{4}$ pizzas = ________ pizzas
7. $4\frac{1}{2}$ pizzas = ________ pizzas
8. $3\frac{5}{6}$ pizzas = ________ pizzas
9. $5\frac{5}{8}$ pizzas = ________ pizzas
10. $2\frac{7}{9}$ pizzas = ________ pizzas
11. $3\frac{1}{8}$ pizzas = ________ pizzas
12. $6\frac{1}{6}$ pizzas = ________ pizzas

Coin Count

There are 100 coins on this page. WRITE the fraction, decimal, and percent for each type of coin.

HINT: A percent (%) is a way of showing parts of 100. If 67 out of 100 coins are pennies, this can be written as $\frac{67}{100}$, 0.67, or 67%.

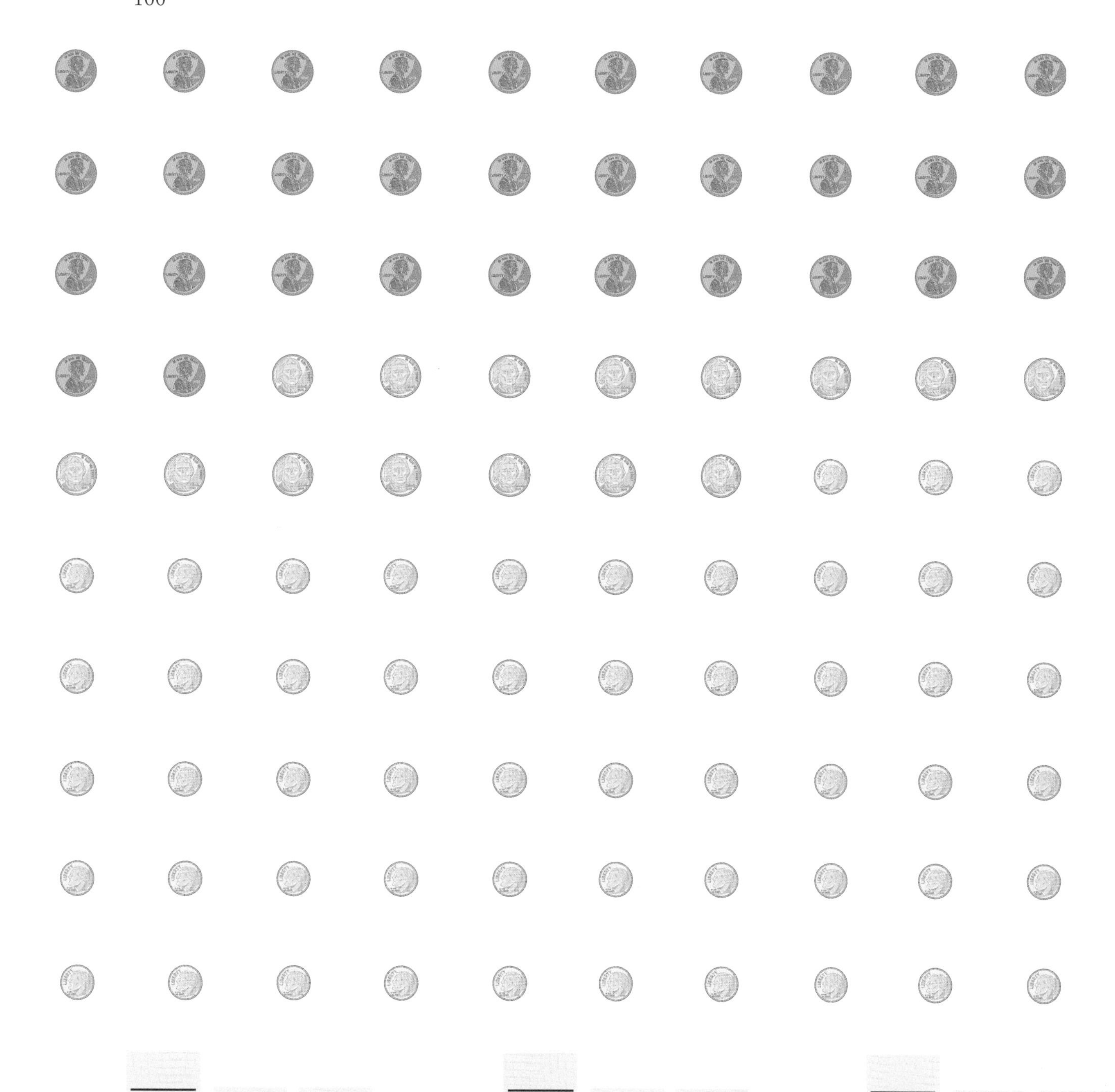

1. Penny: ____ ____ ____ %
2. Nickel: ____ ____ ____ %
3. Dime: ____ ____ ____ %

Frosted Cupcakes

WRITE the fraction, decimal, and percent of yellow frosted cupcakes in each row.

Example:

If 3 out of 10 cupcakes have yellow frosting, this can be written as $\frac{3}{10}$, 0.3, or 30%.

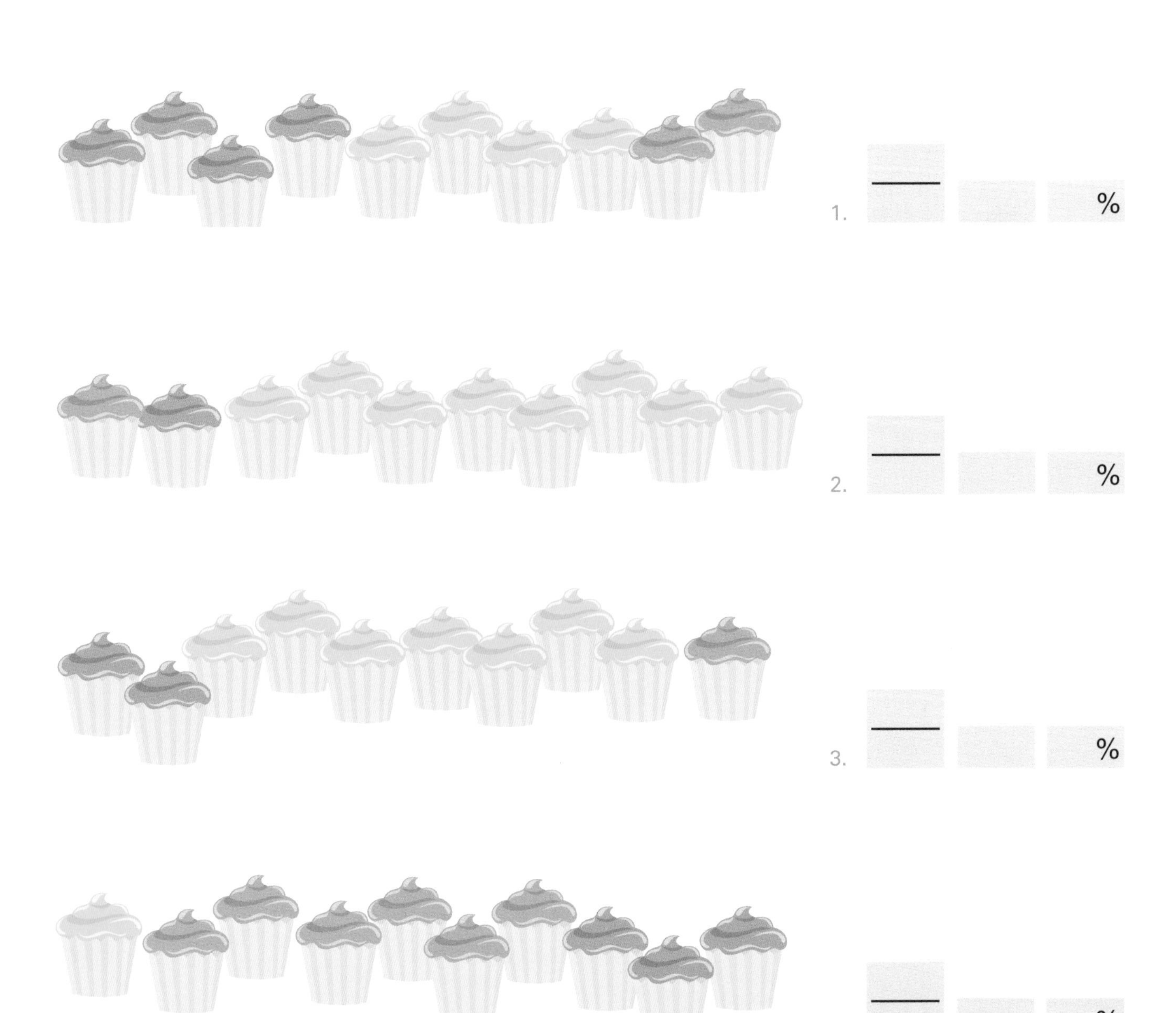

1. ____ ____ %

2. ____ ____ %

3. ____ ____ %

4. ____ ____ %

Survey Says

A research company surveyed people about things that they have in their homes. WRITE the percent for each survey result.

HINT: Think of each group of people as a fraction and the equivalent fraction that has 100 in the denominator.

1. People who own a television set: 99 out of 100 ______%
2. People who own a car: 75 out of 100 ______%
3. People who have at least one pet: 40 out of 50 ______%
4. People who have more than one computer: 6 out of 10 ______%
5. People who own a snowblower: 47 out of 100 ______%
6. People who have a yard with a swing set: 15 out of 50 ______%
7. People who own more than 50 books: 12 out of 25 ______%
8. People who own a motorcycle: 2 out of 100 ______%
9. People who have a home with three bedrooms: 5 out of 10 ______%
10. People who own a boat: 1 out of 25 ______%
11. People who have Internet access: 46 out of 50 ______%
12. People who have a pool: 2 out of 10 ______%

Skateboard Sort

WRITE each percent.

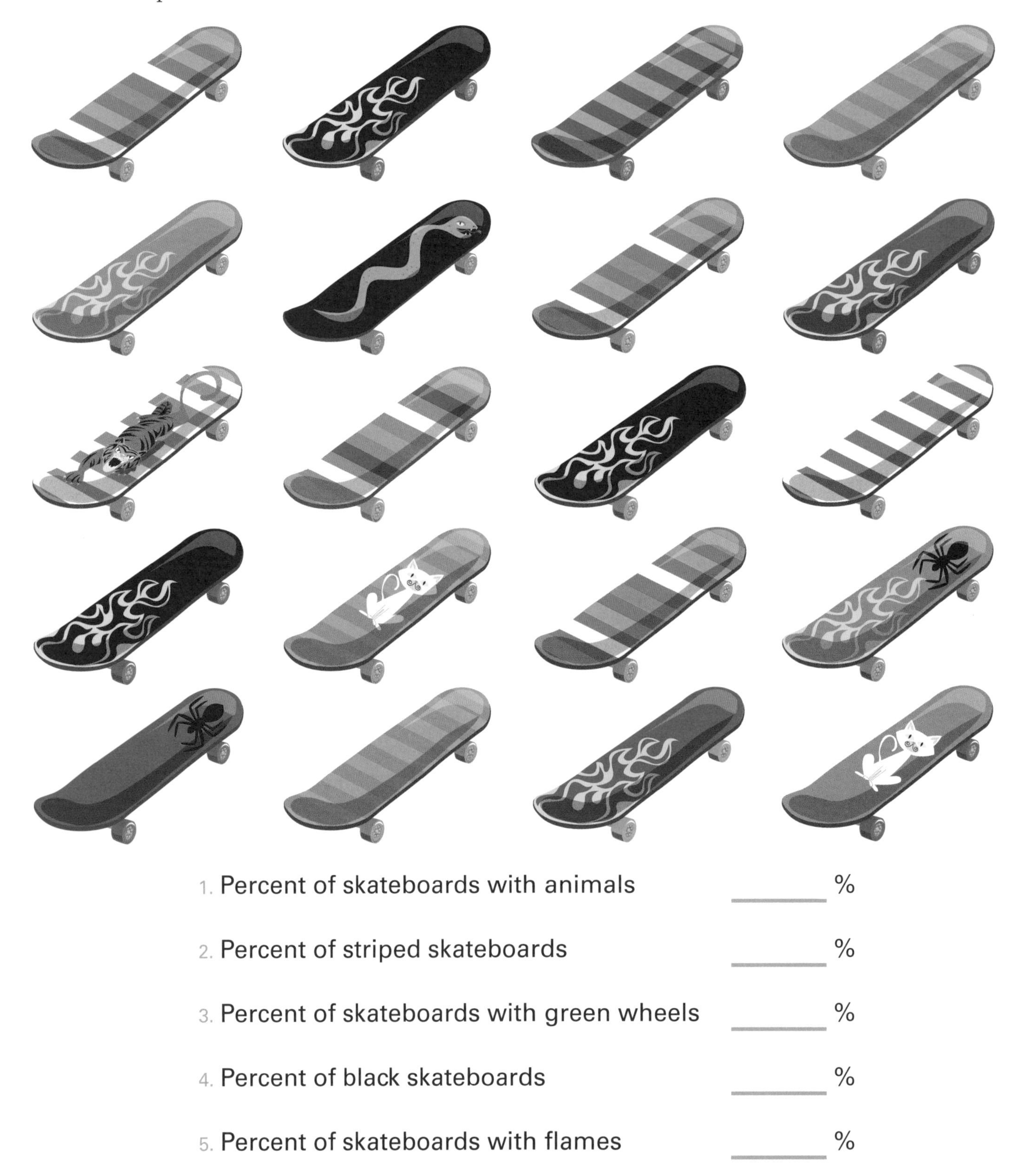

1. Percent of skateboards with animals ________ %
2. Percent of striped skateboards ________ %
3. Percent of skateboards with green wheels ________ %
4. Percent of black skateboards ________ %
5. Percent of skateboards with flames ________ %

Pay the Check

WRITE each tip, and ADD the tip to each of the restaurant checks.

GUEST CHECK

Date	Table	Guests	Server	000945

Hamburger	$4.50
French fries	$1.75
Soda	$1.55
Food Total	$7.80
15% tip	$1.17
Total	$8.97

15% is the same as 0.15, so to find the tip, multiply 7.85 by 0.15.

Then add the tip to the food total, and you have the total cost of the check.

GUEST CHECK

Date	Table	Guests	Server	000946

Meatball sub	$ 6.20
Bag of chips	$ 1.65
Bottle of water	$ 2.35
Food Total	$10.20
10% tip	
Total	

GUEST CHECK

Date	Table	Guests	Server	000947

Chicken club sandwich	$ 8.55
Bowl of soup	$ 3.20
Small coffee	$ 0.85
Food Total	$12.60
15% tip	
Total	

GUEST CHECK

Date	Table	Guests	Server	000948

Grilled ham and cheese	$4.15
Lemonade	$2.40
Cherry pie	$3.30
Food Total	$9.85
20% tip	
Total	

GUEST CHECK

Date	Table	Guests	Server	000949

Garden salad	$10.25
Iced tea	$ 2.60
Chocolate pudding	$ 1.75
Food Total	$14.60
25% tip	
Total	

Cool Collections

READ the descriptions of each collection, and WRITE the answers.

Annie has a collection of 50 action figures. Heroes make up 64% of her action figures, while 36% of her action figures are villains.

1. How many of the action figures are heroes? ________
2. How many of the action figures are villains? ________

Richie has 40 video games. His collection is 50% adventure games, 20% arcade games, and 30% sports games.

3. How many of the video games are adventure games? ________
4. How many of the video games are arcade games? ________
5. How many of the video games are sports games? ________

Kurt has a collection of 125 model train cars. Train engines make up 40% of his collection, while 12% of his collection is cabooses, 16% of his collection is passenger cars, and 32% of his collection is boxcars.

6. How many of the train cars are engines? ________
7. How many of the train cars are cabooses? ________
8. How many of the train cars are passenger cars? ________
9. How many of the train cars are boxcars? ________

Vivian has a collection of 1,500 stamps. Animals appear on 34% of her stamps, pictures of famous people on 52% of her stamps, and flags are featured on 14% of her stamps.

10. How many of the stamps have animals? ________
11. How many of the stamps have people? ________
12. How many of the stamps have flags? ________

Library Checkout

Each person is checking out a large number of library books, all on different topics. What's the largest amount of books on a topic that each person is checking out?
WRITE the fraction, percent, or decimal.

1. Rita checked out 10 books. Of the books Rita borrowed, $\frac{1}{2}$ are about Egypt, 30% are mysteries, and 0.2 are about animals. $\frac{1}{2}$

2. Carl checked out 15 books. Of the books Carl borrowed, $\frac{1}{5}$ are about auto repair, 20% are about robots, and 0.6 have vampires and monsters in them. ______

3. Mei checked out 20 books. Of the books Mei borrowed, $\frac{7}{20}$ are historical fiction, 40% are about Greek mythology, and 0.25 are about wizards. ______

4. Walter checked out 15 books. Of the books Walter borrowed, $\frac{3}{5}$ are biographies, 20% are history books, and 0.2 are about the military. ______

5. Hazel checked out 10 books. Of the books Hazel borrowed, $\frac{2}{5}$ are knitting books, 50% have sewing patterns, and 0.1 is about needlepoint. ______

6. Travis checked out 25 books. Of the books Travis borrowed, $\frac{1}{5}$ are about gardening, 32% are about house repair, and 0.48 are fiction. ______

Best Price

CIRCLE the item with the lower price in each row.

1.

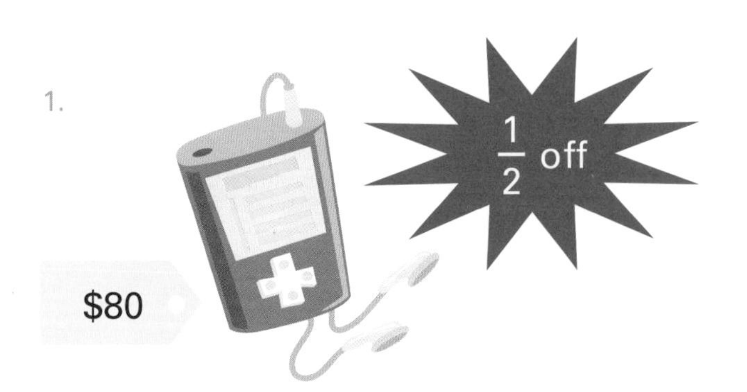

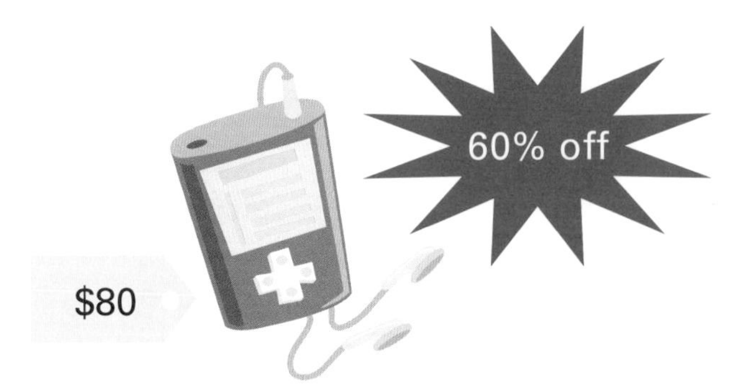

2.

3.

4.

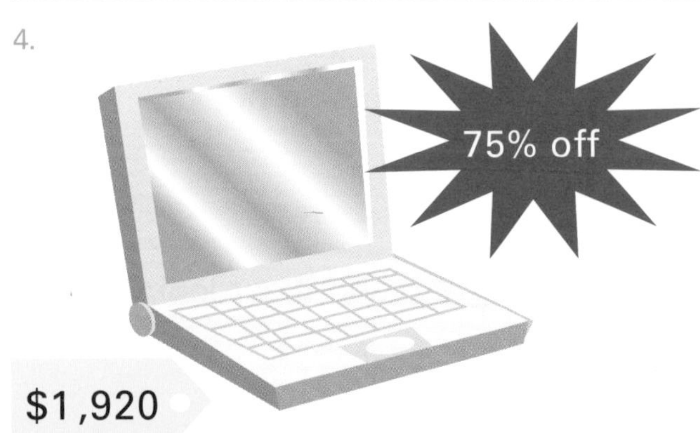

Going Green

Pairs of neighbors are challenging each other to find ways to help the environment. READ the sentences, and WRITE the answers.

On trash day, Robert had $1\frac{1}{4}$ trash cans full and $\frac{9}{8}$ recycling bins full. Sandra had $1\frac{1}{3}$ trash cans full and $\frac{11}{6}$ recycling bins full.

1. If the trash cans and recycling bins are all the same size, who had more recycling than trash? ____________________

Both Alex and Elizabeth have started composting as a way to recycle their food waste instead of throwing it in the trash. Alex has reduced his trash by 35%, and Elizabeth has reduced her trash by $\frac{1}{3}$.

2. Who has a bigger reduction in trash? ____________________

Melissa and Douglas are trying to save on electricity by turning off anything that's not being used. Melissa's electric bill went down by 22%, and Douglas saw his electric bill go down by 0.18.

3. Who's electric bill went down by a larger percent? ____________________

Marcus and Amaryllis are both trying to walk and bike more than ride in a car wherever they go. In the past month, Amaryllis's family has bought $4\frac{3}{8}$ fewer gallons of gasoline than usual, and Marcus's family has bought $\frac{23}{5}$ fewer gallons of gasoline than usual.

4. Whose family had the bigger gasoline savings? ____________________

Best Price

WRITE the price of each video game console. Then CIRCLE the one with the lowest price.

Leftover Pizza

What fraction of pizza can be made if two people combine their leftover pizza? WRITE the fraction in its simplest form.

1.

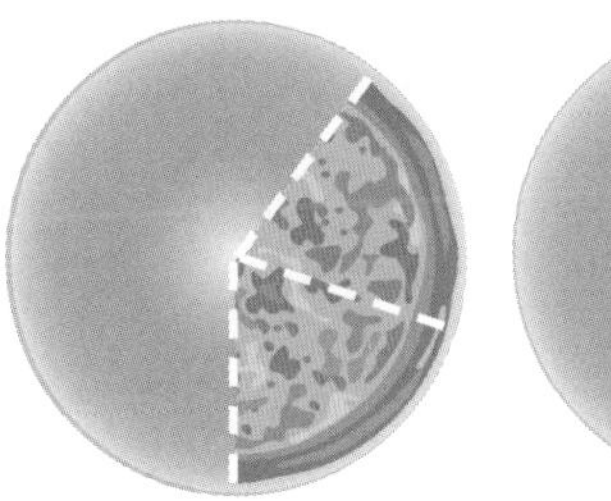

$\frac{2}{5} + \frac{1}{5} =$ ____

2.

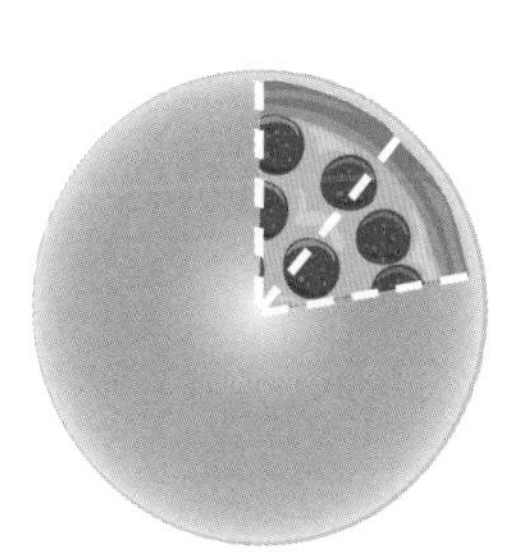
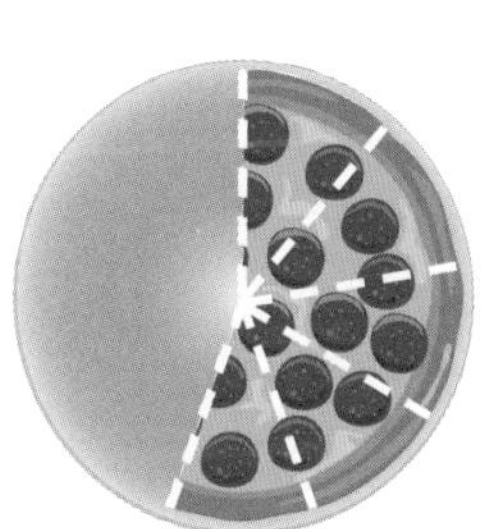

$\frac{2}{9} + \frac{5}{9} =$ ____

3.

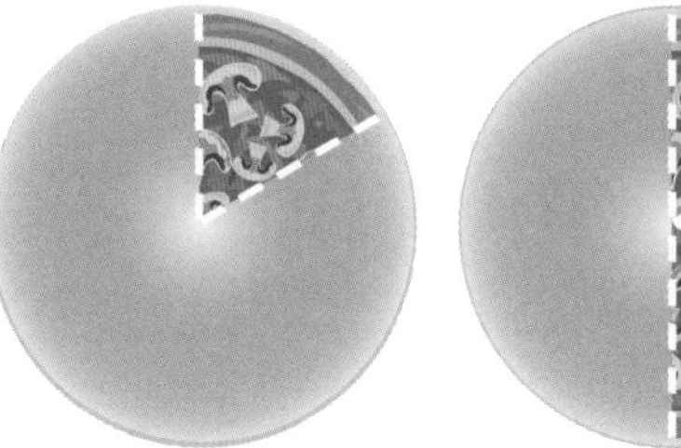

$\frac{1}{6} + \frac{3}{6} =$ ____

4.

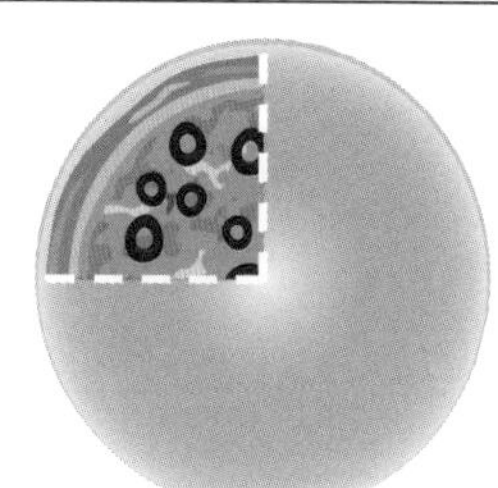
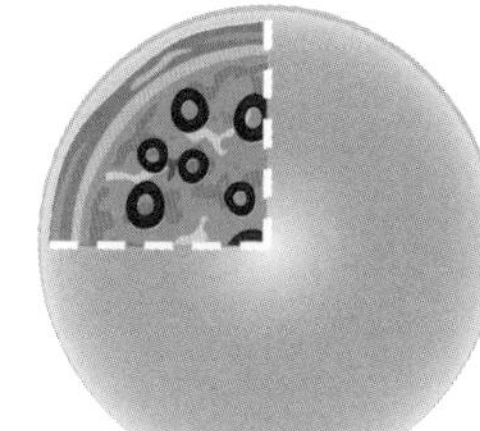

$\frac{1}{4} + \frac{1}{4} =$ ____

5.

$\frac{1}{10} + \frac{5}{10} =$ ____

6.

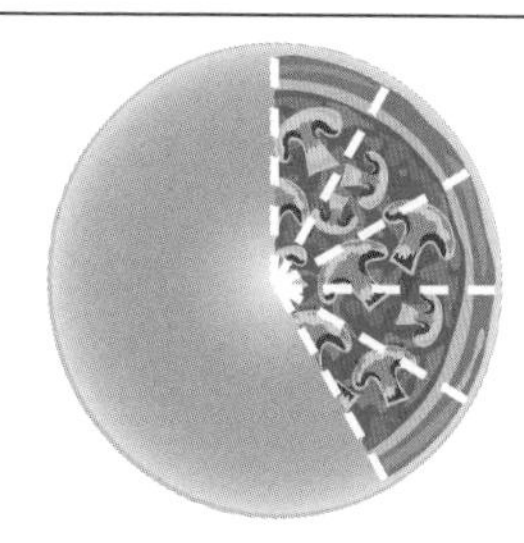

$\frac{5}{12} + \frac{5}{12} =$ ____

7.

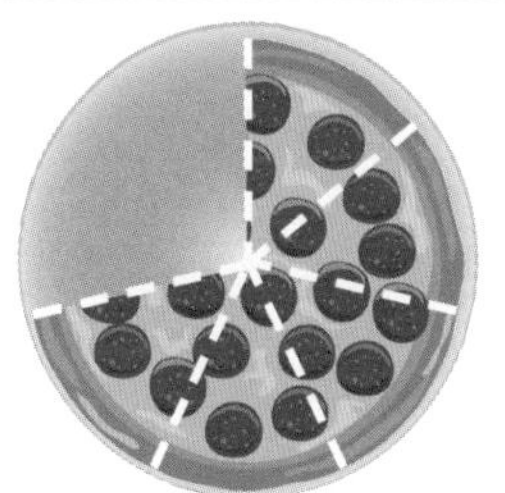

$\frac{5}{7} + \frac{3}{7} =$ ____

8.

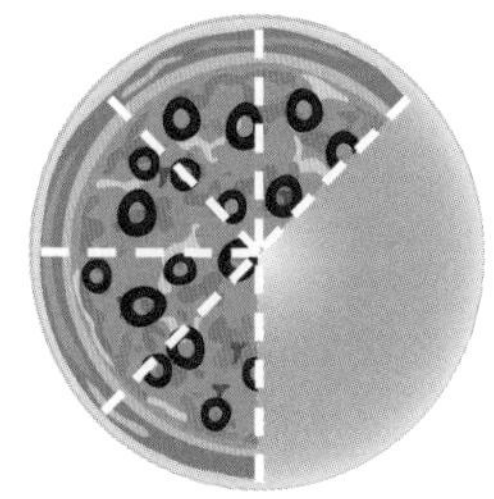
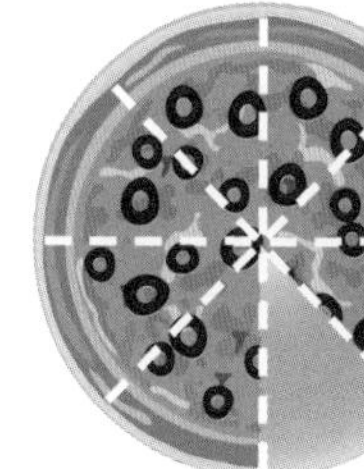

$\frac{5}{8} + \frac{7}{8} =$ ____

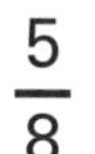

Kate's Kitchen

Today in Kate's Kitchen, Kate is making two different kinds of brownies. FIND the total amount of each ingredient that will be needed to make both recipes, and WRITE the fraction in its simplest form.

Scrumptious Fudge Brownies	Cakey Chocolate Brownies	Ingredients
$\frac{1}{2}$ cup butter	$\frac{3}{4}$ cup butter	1. ____ cup butter
$\frac{3}{4}$ cup sugar	$\frac{2}{3}$ cup sugar	2. ____ cup sugar
$\frac{1}{3}$ cup cocoa powder	$\frac{1}{2}$ cup cocoa powder	3. ____ cup cocoa powder
$\frac{3}{4}$ cup flour	$\frac{1}{3}$ cup flour	4. ____ cup flour
$\frac{1}{4}$ cup heavy cream	$\frac{1}{3}$ cup heavy cream	5. ____ cup heavy cream
$\frac{1}{4}$ teaspoon baking powder	$\frac{1}{2}$ teaspoon baking powder	6. ____ teaspoon baking powder

That Does Not Compute!

The Great Roboto is on the fritz and is spitting out some math problems with wrong answers. CIRCLE the incorrect sums.

$\frac{5}{8} + \frac{2}{4} = \frac{7}{8}$

$\frac{4}{9} + \frac{1}{3} = \frac{7}{9}$

$\frac{7}{12} + \frac{3}{8} = 1\frac{1}{24}$

$\frac{1}{2} + \frac{7}{8} = 1\frac{3}{4}$

$\frac{5}{6} + \frac{1}{12} = 1$

$\frac{3}{7} + \frac{4}{21} = \frac{13}{21}$

$\frac{6}{7} + \frac{3}{14} = 1\frac{1}{14}$

$\frac{4}{15} + \frac{2}{3} = \frac{14}{15}$

$\frac{3}{5} + \frac{9}{10} = \frac{2}{3}$

$\frac{7}{9} + \frac{1}{6} = \frac{17}{18}$

$\frac{3}{11} + \frac{17}{22} = 1\frac{1}{22}$

$\frac{8}{9} + \frac{2}{5} = 1\frac{2}{9}$

Gassing Up

How much gasoline will each person have after a stop at the gas station? WRITE the fraction in its simplest form.

1. Burt had $2\frac{3}{5}$ gallons of gas in his car, and he added $8\frac{1}{5}$ gallons at the gas station. ______ gallons

2. Marsha had $6\frac{1}{6}$ gallons of gas in her car, and she added $9\frac{5}{6}$ gallons at the gas station. ______ gallons

3. David had $7\frac{3}{4}$ gallons of gas in his SUV, and he added $14\frac{1}{2}$ gallons at the gas station. ______ gallons

4. Teri had $1\frac{3}{8}$ gallons of gas in her car, and she added $12\frac{1}{4}$ gallons at the gas station. ______ gallons

5. Michael had $5\frac{9}{14}$ gallons of gas in his car, and he added $11\frac{4}{7}$ gallons at the gas station. ______ gallons

6. Rebecca had $6\frac{1}{5}$ gallons of gas in her truck, and she added $18\frac{3}{10}$ gallons at the gas station. ______ gallons

7. Peter had $4\frac{5}{6}$ gallons of gas in his van, and he added $14\frac{1}{3}$ gallons at the gas station. ______ gallons

8. Janet had $12\frac{4}{9}$ gallons of gas in her SUV, and she added $16\frac{2}{3}$ gallons at the gas station. ______ gallons

Louisa's Lasagna

Each lasagna has been cut into a different number of pieces, and some of the pieces have been eaten. WRITE the remaining fraction of lasagna as a fraction in its simplest form.

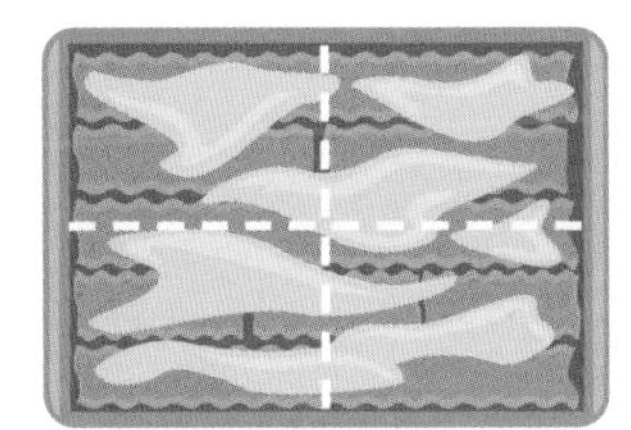

1. Louisa cut her lasagna into 4 pieces, and 2 pieces were eaten.

$$\frac{4}{4} - \frac{2}{4} = \underline{\quad\quad}$$

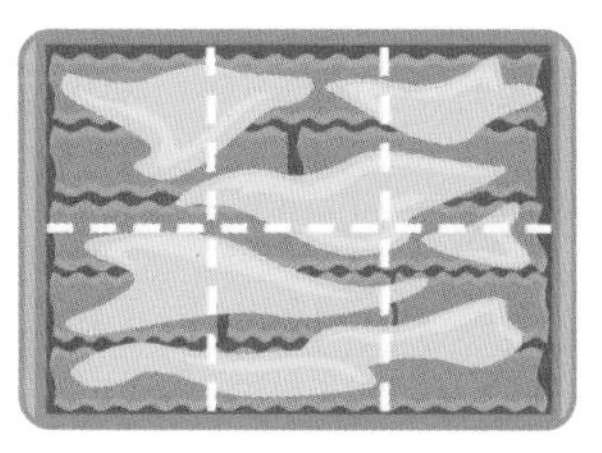

2. Louisa cut her lasagna into 6 pieces, and 4 pieces were eaten.

$$\frac{6}{6} - \frac{4}{6} = \underline{\quad\quad}$$

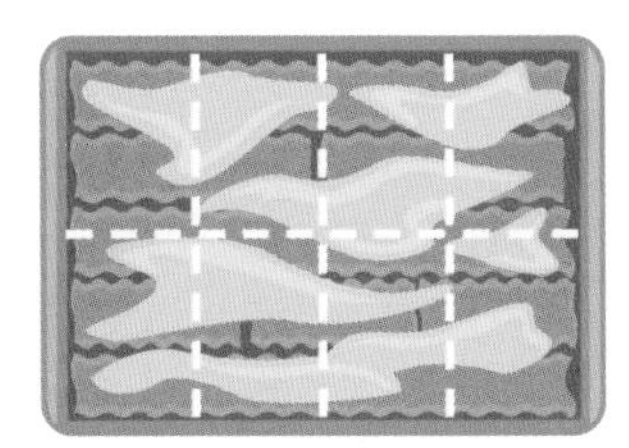

3. Louisa cut her lasagna into 8 pieces, and 5 pieces were eaten.

$$\frac{8}{8} - \frac{5}{8} = \underline{\quad\quad}$$

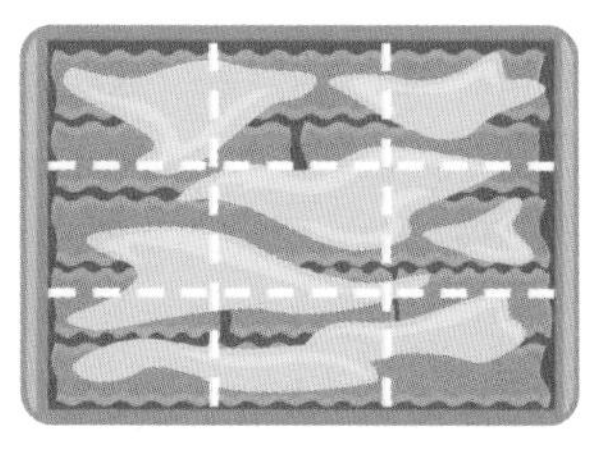

4. Louisa cut her lasagna into 9 pieces, and 3 pieces were eaten.

$$\frac{9}{9} - \frac{3}{9} = \underline{\quad\quad}$$

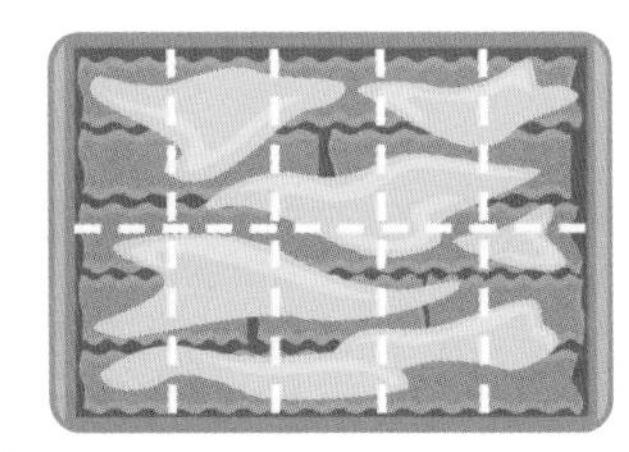

5. Louisa cut her lasagna into 10 pieces, and 8 pieces were eaten.

$$\frac{10}{10} - \frac{8}{10} = \underline{\quad\quad}$$

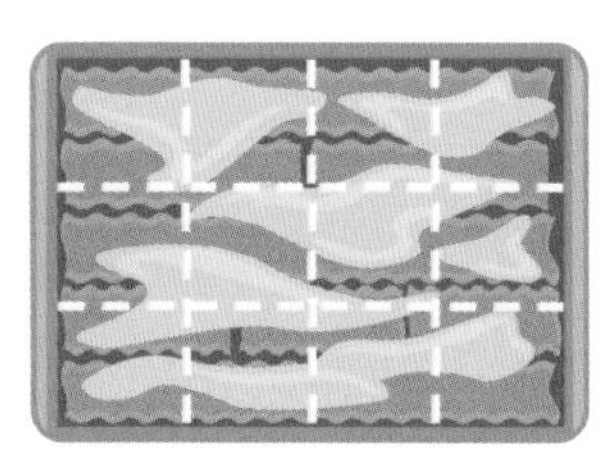

6. Louisa cut her lasagna into 12 pieces, and 4 pieces were eaten.

$$\frac{12}{12} - \frac{4}{12} = \underline{\quad\quad}$$

Rainy Days

The Jones family has a rain gauge in the backyard to measure the amount of rain that falls each day. The chart shows the amount of rain that fell in one rainy week in April. LOOK at the chart, and WRITE each answer as a fraction in its simplest form.

Sunday	Monday	Tuesday	Wednesday	Thursday	Friday	Saturday
$\frac{3}{4}$ in.	$\frac{1}{3}$ in.	$\frac{3}{8}$ in.	$\frac{1}{4}$ in.	$1\frac{1}{2}$ in.	$\frac{7}{8}$ in.	$\frac{3}{5}$ in.

1. How much more rain fell on Friday than on Tuesday? _____ in.
2. How much more rain fell on Sunday than on Wednesday? _____ in.
3. How much more rain fell on Thursday than on Sunday? _____ in.
4. How much more rain fell on Monday than on Wednesday? _____ in.
5. How much more rain fell on Sunday than on Saturday? _____ in.
6. How much more rain fell on Thursday than on Friday? _____ in.
7. How much more rain fell on Saturday than on Monday? _____ in.
8. How much more rain fell on Friday than on Saturday? _____ in.

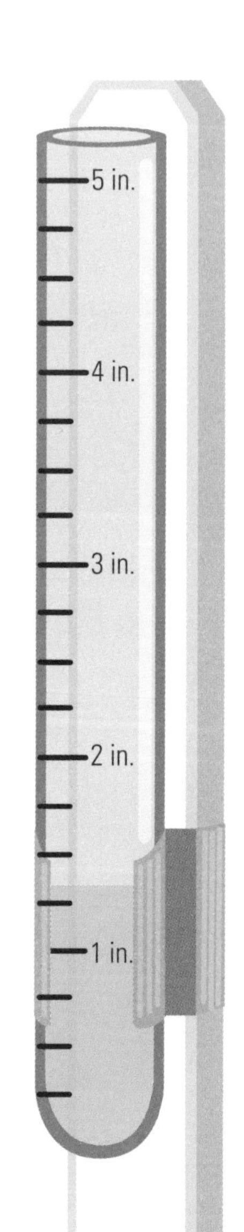

That Does Not Compute!

The Great Roboto is on the fritz and is spitting out some math problems with wrong answers. CIRCLE the incorrect differences.

$\frac{5}{6} - \frac{5}{18} = \frac{5}{9}$

$\frac{7}{9} - \frac{2}{3} = \frac{1}{9}$

$\frac{11}{12} - \frac{1}{4} = \frac{7}{12}$

$\frac{3}{4} - \frac{1}{8} = \frac{5}{8}$

$\frac{5}{6} - \frac{3}{5} = \frac{2}{6}$

$\frac{3}{4} - \frac{2}{3} = \frac{1}{12}$

$\frac{9}{14} - \frac{2}{7} = \frac{5}{14}$

$\frac{7}{15} - \frac{1}{5} = \frac{2}{5}$

$\frac{7}{10} - \frac{1}{2} = \frac{1}{10}$

$\frac{8}{9} - \frac{5}{6} = \frac{1}{36}$

$\frac{15}{16} - \frac{3}{4} = \frac{1}{4}$

$\frac{5}{8} - \frac{1}{24} = \frac{7}{12}$

Tall Team

READ the paragraph, and WRITE each answer as a fraction in its simplest form.

The Barnaby Basketball team is lucky to have so many tall players. The tallest player, Marco, is $76\frac{3}{8}$ inches tall. The shortest player on the team is Lucy, who is $68\frac{1}{4}$ inches tall. Henry is $72\frac{1}{3}$ inches tall, and Tamara is close to Henry's height at $71\frac{3}{4}$ inches tall. The last player is Jack, who is $70\frac{1}{2}$ inches tall.

1. How much taller is Tamara than Lucy? ______ in.
2. How much taller is Marco than Lucy? ______ in.
3. How much taller is Tamara than Jack? ______ in.
4. How much taller is Henry than Lucy? ______ in.
5. How much taller is Marco than Jack? ______ in.
6. How much taller is Henry than Tamara? ______ in.

Time Well Spent

READ the paragraph, and WRITE each answer as a fraction in its simplest form.

Malcolm follows the same schedule after school every Monday through Friday. He spends $\frac{1}{4}$ hour on the bus to karate class, which lasts $\frac{3}{4}$ hour. His mom picks him up and drives him home, which takes $\frac{1}{3}$ hour. At home he has his $\frac{1}{2}$-hour-long violin lesson, followed by $\frac{2}{3}$ hour of computer time.

Then he relaxes with $\frac{5}{6}$ hour of television.

What is the total amount of time Malcolm spends doing each activity in a five-day period?

1. Riding the bus _____ hours
2. Taking a karate class _____ hours
3. Riding in the car _____ hours
4. Taking a violin lesson _____ hours
5. Using the computer _____ hours
6. Watching television _____ hours

Kate's Kitchen

Today in Kate's Kitchen, Kate is scaling up a recipe. When you scale up a recipe, you multiply the ingredients in the recipe to make a larger number of servings. WRITE the new amount of each ingredient in the scaled-up recipes as a fraction in its simplest form.

HINT: Multiply the ingredients by 2 and by 4 to scale up the recipes.

Very Berry Muffins Serves 6	Very Berry Muffins Serves 12	Very Berry Muffins Serves 24
2 eggs	1. 4 eggs	9. 8 eggs
$\frac{5}{8}$ stick of butter	2. ______ stick of butter	10. ______ stick of butter
$\frac{3}{4}$ cup sugar	3. ______ cup sugar	11. ______ cup sugar
$\frac{1}{4}$ cup brown sugar	4. ______ cup brown sugar	12. ______ cup brown sugar
$1\frac{1}{2}$ cups flour	5. ______ cups flour	13. ______ cups flour
$\frac{1}{2}$ teaspoon baking powder	6. ______ teaspoon baking powder	14. ______ teaspoon baking powder
$\frac{4}{5}$ cup blueberries	7. ______ cup blueberries	15. ______ cup blueberries
$\frac{1}{3}$ cup raspberries	8. ______ cup raspberries	16. ______ cup raspberries

That Does Not Compute!

The Great Roboto is on the fritz and is spitting out some math problems with wrong answers. CIRCLE the incorrect products.

$\frac{4}{5} \times \frac{1}{16} = \frac{2}{5}$

$\frac{1}{5} \times \frac{3}{4} = \frac{1}{10}$

$\frac{7}{8} \times \frac{1}{14} = \frac{1}{16}$

$\frac{1}{20} \times \frac{5}{6} = \frac{1}{24}$

$\frac{1}{2} \times \frac{6}{7} = \frac{3}{7}$

$\frac{5}{28} \times \frac{7}{15} = \frac{5}{6}$

$\frac{3}{7} \times \frac{14}{15} = \frac{2}{5}$

$\frac{7}{12} \times \frac{6}{7} = \frac{1}{7}$

$\frac{5}{9} \times \frac{1}{5} = \frac{1}{9}$

$\frac{16}{17} \times \frac{15}{16} = \frac{17}{15}$

$\frac{2}{3} \times \frac{33}{34} = \frac{11}{17}$

$\frac{3}{11} \times \frac{7}{12} = \frac{7}{12}$

Tough Training

Maxine is training to run a marathon. With each part of her training, she runs longer distances for a fewer number of days. WRITE the total number of miles Maxine will run in each part of the training.

Part	Days	Miles	Total Miles
1	40	$2\frac{1}{2}$	100
2	35	$4\frac{1}{5}$	
3	30	$6\frac{3}{5}$	
4	24	$10\frac{2}{3}$	
5	18	$15\frac{5}{6}$	
6	12	$18\frac{3}{4}$	
7	7	$22\frac{4}{7}$	
8	5	$26\frac{2}{5}$	

Time Slots

WRITE each answer.

1. How many $\frac{1}{2}$ hours are in 24 hours? 48
2. How many $\frac{1}{3}$ hours are in 6 hours? ______
3. How many $\frac{1}{4}$ hours are in 48 hours? ______
4. How many $\frac{3}{4}$ hours are in 30 hours? ______
5. How many $\frac{5}{6}$ hours are in 15 hours? ______
6. How many $\frac{1}{4}$ hours are in $\frac{3}{4}$ hour? ______
7. How many $\frac{1}{2}$ hours are in $\frac{5}{2}$ hours? ______
8. How many $\frac{1}{9}$ hours are in $\frac{2}{3}$ hour? ______
9. How many $\frac{1}{6}$ hours are in $\frac{1}{2}$ hour? ______
10. How many $\frac{1}{12}$ hours are in $\frac{3}{4}$ hour? ______

Part of a Pie

WRITE the size of each piece of pie as a fraction in its simplest form.

1. Pete is cutting $\frac{3}{4}$ of a pie into 6 pieces.
   What size will each piece be? $\frac{1}{8}$ of the whole pie

2. Betsy is cutting $\frac{4}{5}$ of a pie into 8 pieces.
   What size will each piece be? ____ of the whole pie

3. Don is cutting $\frac{1}{2}$ of a pie into 7 pieces.
   What size will each piece be? ____ of the whole pie

4. Joan is cutting $\frac{7}{8}$ of a pie into 7 pieces.
   What size will each piece be? ____ of the whole pie

5. Roger is cutting $\frac{8}{9}$ of a pie into 4 pieces.
   What size will each piece be? ____ of the whole pie

6. Peggy is cutting $\frac{2}{3}$ of a pie into 10 pieces.
   What size will each piece be? ____ of the whole pie

7. Kenny is cutting $\frac{9}{10}$ of a pie into 6 pieces.
   What size will each piece be? ____ of the whole pie

8. Trudy is cutting $\frac{6}{7}$ of a pie into 3 pieces.
   What size will each piece be? ____ of the whole pie

That Does Not Compute!

The Great Roboto is on the fritz and is spitting out some math problems with wrong answers. CIRCLE the incorrect quotients.

$\frac{1}{9} \div \frac{2}{3} = \frac{1}{6}$

$\frac{8}{13} \div \frac{3}{26} = 5\frac{1}{3}$

$\frac{5}{9} \div \frac{5}{27} = 3$

$\frac{7}{8} \div \frac{1}{16} = \frac{1}{14}$

$\frac{2}{5} \div \frac{3}{8} = \frac{15}{16}$

$1 \div \frac{4}{5} = \frac{4}{5}$

$\frac{3}{11} \div \frac{5}{21} = \frac{7}{55}$

$\frac{4}{7} \div \frac{8}{9} = \frac{9}{14}$

$\frac{5}{3} \div \frac{5}{3} = \frac{6}{25}$

$\frac{9}{10} \div \frac{1}{40} = \frac{1}{36}$

$\frac{7}{8} \div \frac{21}{24} = 1$

$\frac{6}{7} \div \frac{12}{35} = 2\frac{1}{2}$

Super Stacker

Lela has been busy stacking things she finds around the house. WRITE the number of objects in each stack.

1. Lela made a stack of nickels that is $5\frac{7}{8}$ inches tall. Each nickel is $\frac{1}{16}$ inches tall. How many coins are in the stack? ______

2. Lela has blocks that are $1\frac{3}{8}$ inches tall, and with her blocks she made a stack that is $20\frac{5}{8}$ inches tall. How many blocks are in the stack? ______

3. Lela used cookies that are $\frac{7}{16}$ inches tall to make a stack of cookies that is $9\frac{5}{8}$ inches tall. How many cookies are in the stack? ______

4. Lela made a stack of cups that measures $40\frac{5}{8}$ inches tall. Each cup is $3\frac{1}{8}$ inches tall. How many cups are in the stack? ______

5. Lela made a stack of sugar cubes measuring $21\frac{1}{4}$ inches from sugar cubes that are each $\frac{5}{8}$ inches tall. How many sugar cubes are in the stack? ______

6. Lela used magnets that are $\frac{5}{16}$ inches tall to make a stack of magnets that is $17\frac{1}{2}$ inches tall. How many magnets are in the stack? ______

Kate's Kitchen

Today in Kate's Kitchen, Kate is making trail mix. WRITE the total number of cups of trail mix that Kate is making. Then WRITE the number of $\frac{1}{4}$ cup servings that can be made from the trail mix.

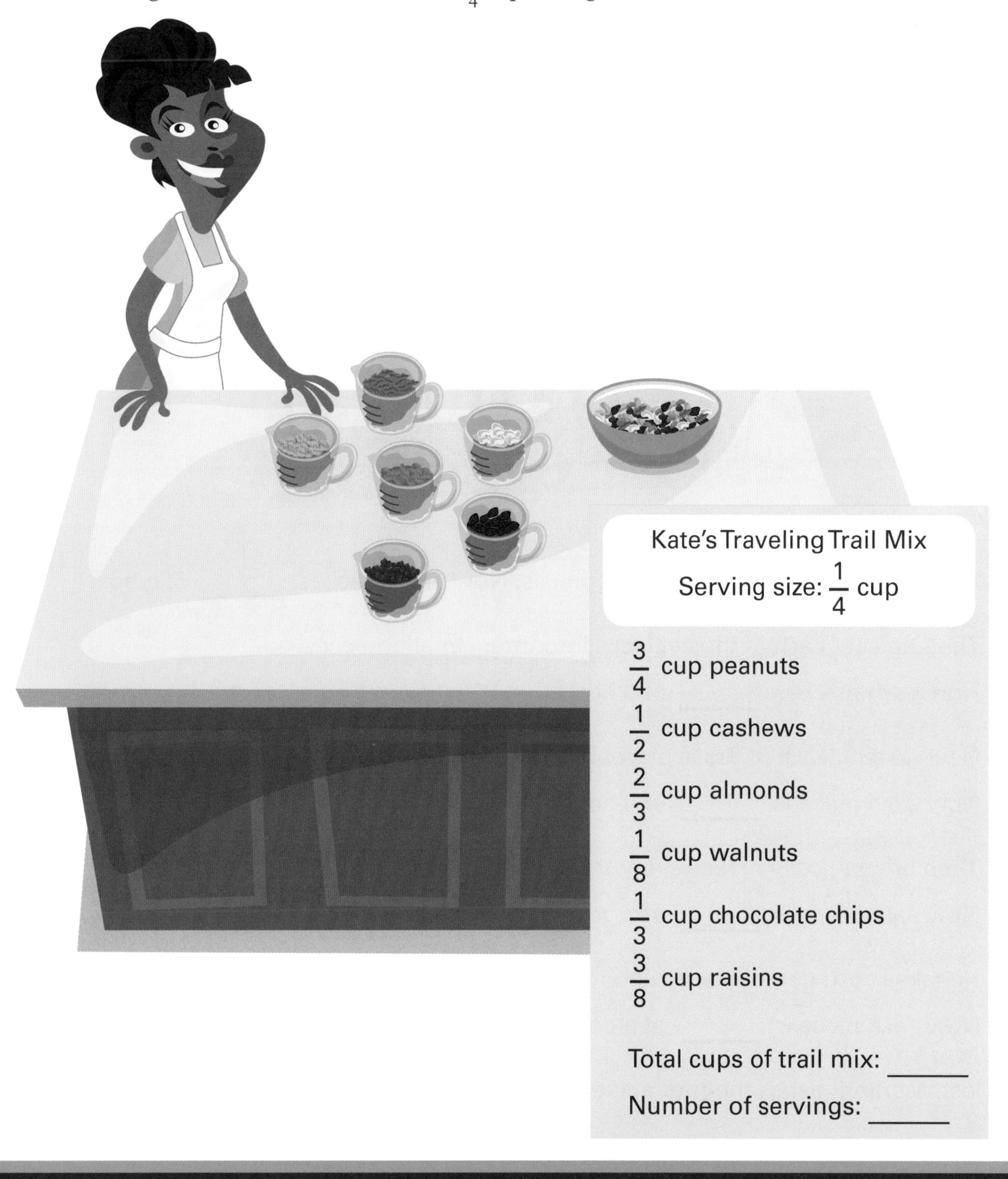

Magic Trick

Merlin the Magnifico does a trick where he saws a box into pieces and puts the pieces back together in different ways. WRITE the size of each box piece for each step in the magic trick.

1. First Merlin cuts the box into two pieces. Now each piece is $\frac{1}{2}$ of a box.
2. Then he cuts each of those pieces into three smaller pieces.
   Now each piece is ______ of a box.
3. Next he cuts each of those pieces into four smaller pieces.
   Now each piece is ______ of a box.
4. Then he magically makes every three pieces into one piece.
   Now each piece is ______ of a box.
5. He takes each pair of pieces and magically transforms them into one piece.
   Now each piece is ______ of a box.

At last, Merlin finishes the trick and transforms the pieces back into one box. Ta-da!

Lightning Chargers

Gary is really proud of his Lightning Charger action figures. He never takes any of them out of their boxes, and he wants to build shelves to show off his collection. MEASURE the width of the box in inches. Then, ANSWER the questions. Write the answers as fractions.

HINT: Remember, 1 foot (ft) = 12 inches (in.), and 1 yard (yd) = 36 inches.

Gary has 57 of the Lightning Charger action figures that all come in the same size box. He wants to group them in different ways. For each of these groups, how long should he make each shelf?

1. Shelf 1 should have 11 action figures. _______ in.
2. Shelf 2 should have 13 action figures. _______ in.
3. Shelf 3 should have 15 action figures. _______ in.
4. Shelf 4 should have 18 action figures. _______ in.
5. If Gary builds only shelves that are one yard long, how many action figures would fit on a shelf?

   ____

6. If Gary builds only shelves that are one foot long, how many action figures would fit on a shelf?

   ____

Alien Invasion

The aliens have landed... and they're tiny!
Scientists want to study the little visitors.
MEASURE each alien. Then ANSWER the questions.

HINT: Remember, 1 meter (m) = 100 centimeters (cm).

What is the width of each alien? Write the answers as decimals.

1. green ________ cm
2. blue ________ cm
3. orange ________ cm

Their alien spaceship is one meter wide. How many of each alien standing side to side would be about as long as the spaceship?

4. green ________ cm
5. blue ________ cm
6. orange ________ cm

Angled Alphabet

There are three different types of angles: right, acute, and obtuse. LOOK at the letters, and WRITE the number of right, acute, and obtuse angles that can be found in each letter.

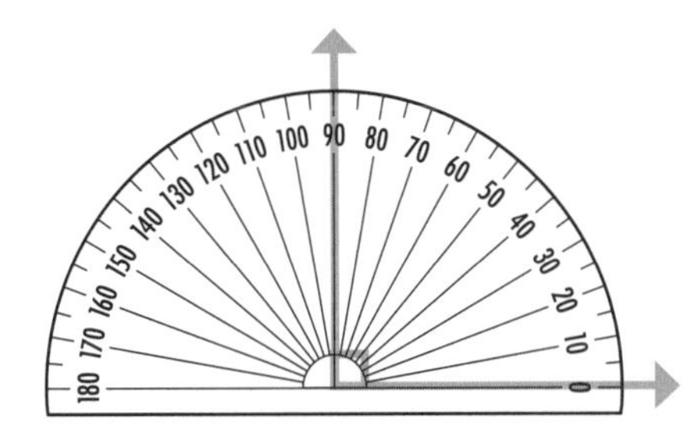

A **right** angle is an angle measuring exactly 90°, indicated by the ⌉ symbol in the corner.

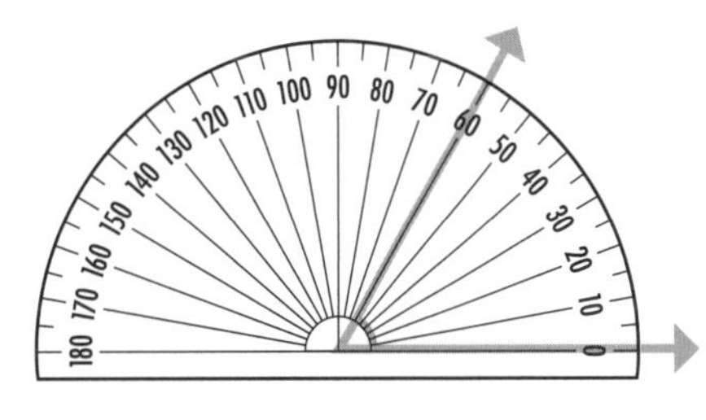

An **acute** angle is any angle measuring less than 90°.

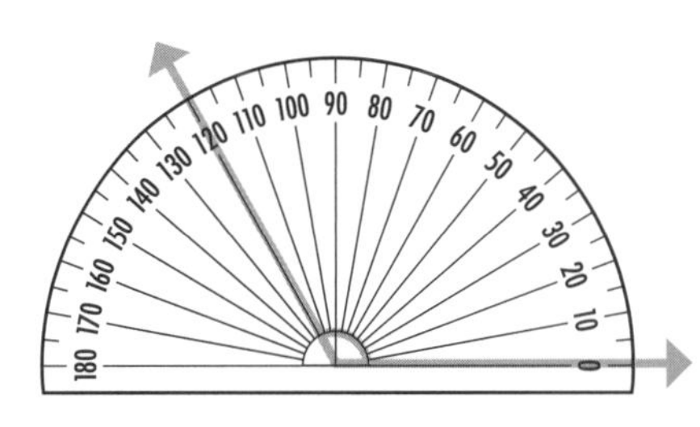

An **obtuse** angle is any angle measuring more than 90°.

1. A
   _____ right
   _____ acute
   _____ obtuse

2. F
   _____ right
   _____ acute
   _____ obtuse

3. H
   _____ right
   _____ acute
   _____ obtuse

4. X
   _____ right
   _____ acute
   _____ obtuse

5. W
   _____ right
   _____ acute
   _____ obtuse

6. Y
   _____ right
   _____ acute
   _____ obtuse

Angle Creator

DRAW an angle to match each measurement.

HINT: If you don't have a protractor, cut out the one in the example.

Example:

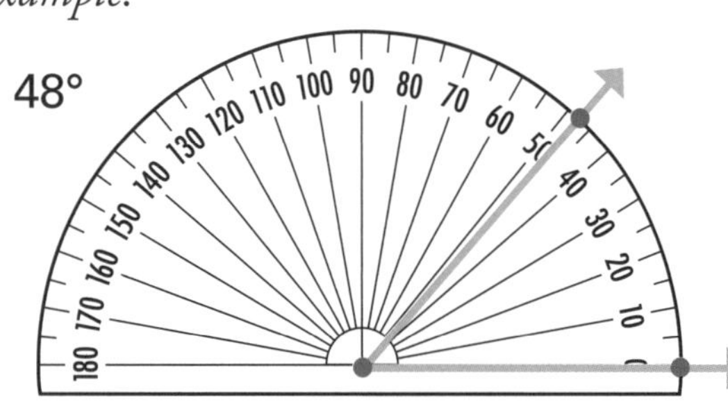

Mark a point on the page. Place the center of the protractor on the point. Place a second point on the paper at zero on the protractor. Place a third point on the paper at the measurement. Then connect the first point to the second point. Finally connect the first point to the third point to form the angle.

1. 15°

2. 36°

3. 75°

4. 142°

5. 120°

6. 94°

Line Art

Each person was asked to draw the following: A, $\overline{BC}$, $\overleftrightarrow{DE}$, and $\overrightarrow{FG}$. CIRCLE the paper with the correct drawings.

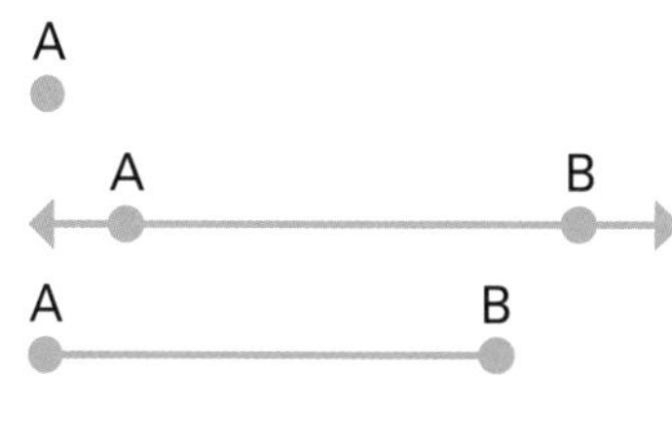

A is a **point**. A point marks a place in space, represented by a dot.

$\overleftrightarrow{AB}$ is a **line**. A line is a straight path that has no end in either direction.

$\overline{AB}$ is a **line segment**. A line segment is the part of a line between two points, called **endpoints**.

$\overrightarrow{AB}$ is a **ray**. A ray is a line that begins at an endpoint and has no end in the other direction.

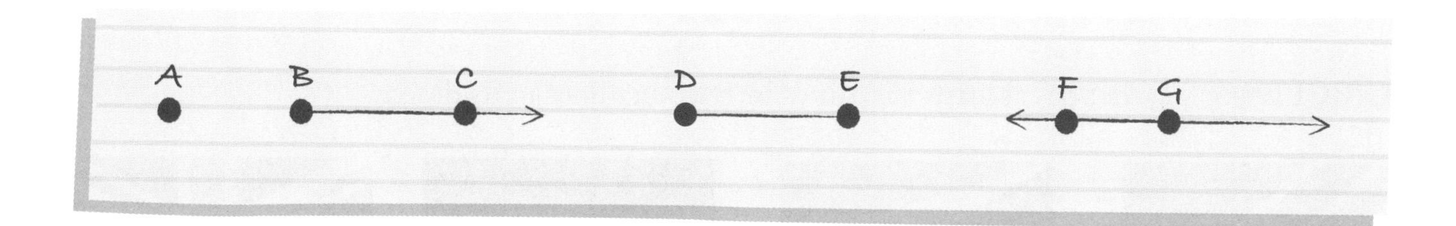

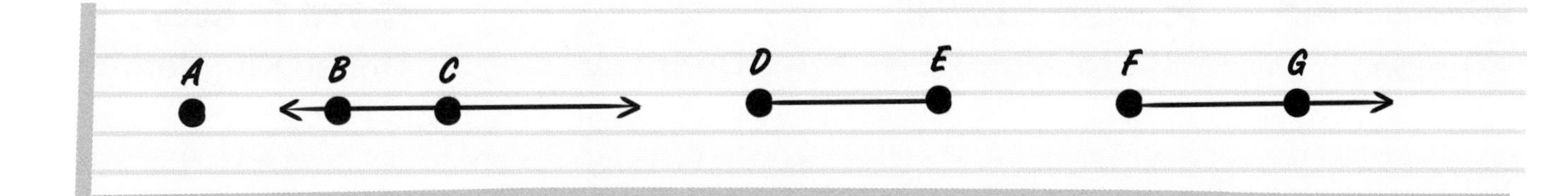

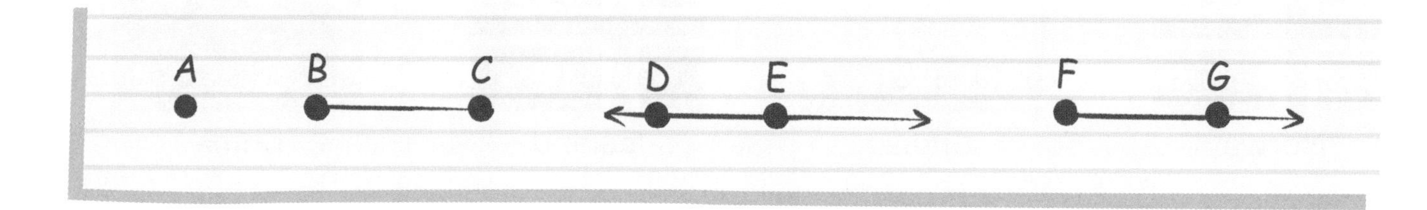

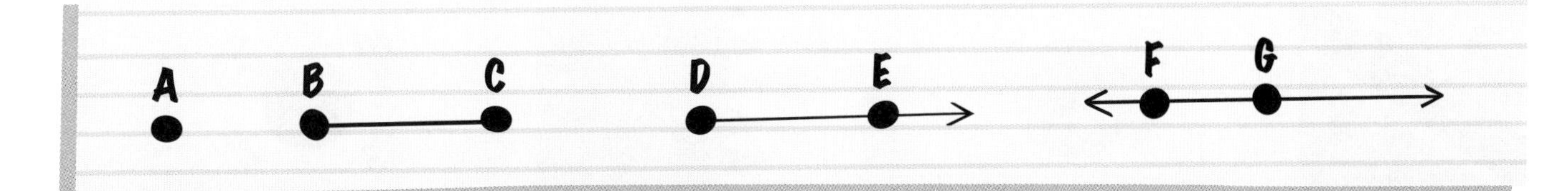

Find the Flag

CIRCLE the flags that match each description.

HINT: Do not count the flag edges. Only use the flag pattern.

Intersecting lines are lines that cross one another.

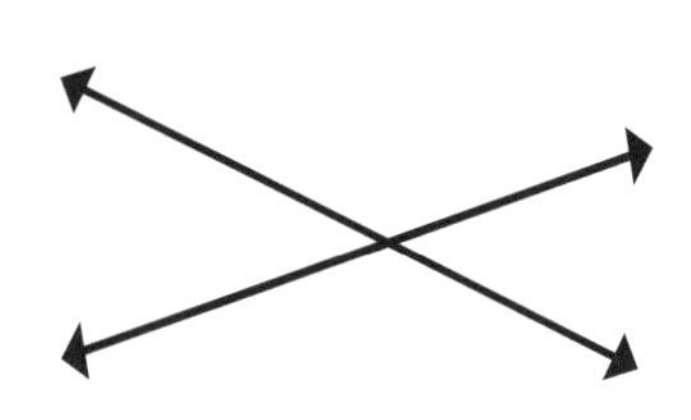

Perpendicular lines intersect to form right angles.

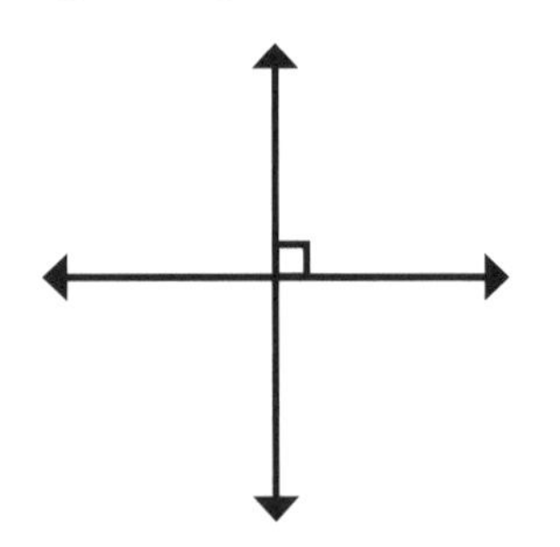

Parallel lines never intersect and are always the same distance apart.

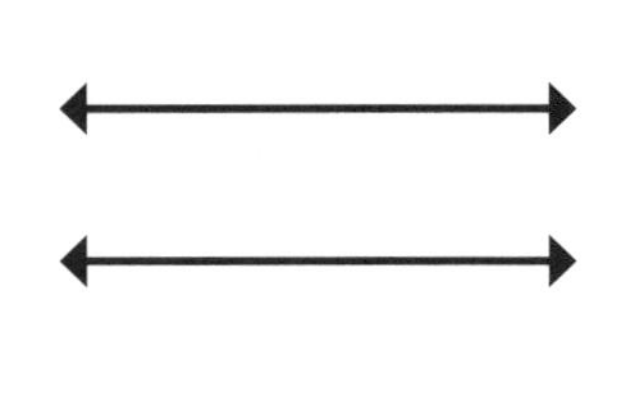

CIRCLE any flag that has at least one pair of intersecting lines in its design.

Trinidad

Guyana

Germany

United Kingdom

CIRCLE any flag that has at least one pair of perpendicular lines in its design.

Norway

India

Iceland

China

CIRCLE any flag that has at least one pair of parallel lines in its design.

Tanzania

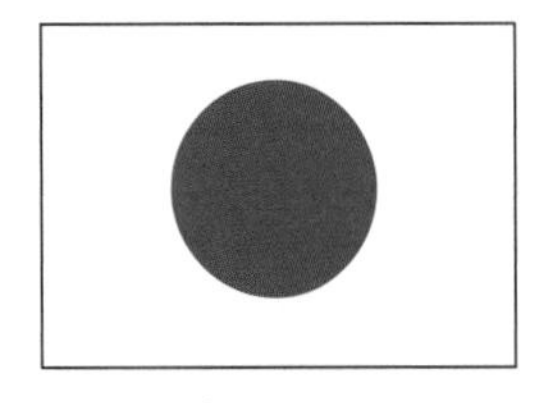

Japan

Bahrain

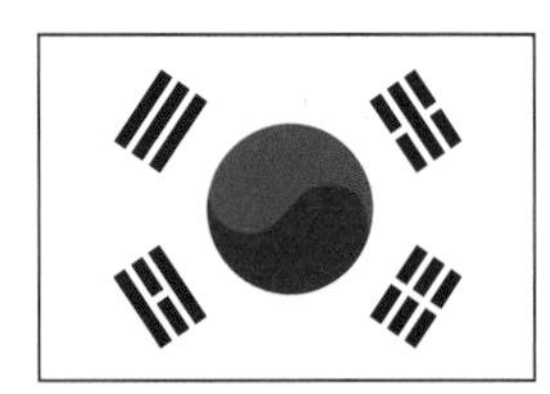

South Korea

Making Polygons

A polygon is a closed plane shape that has three or more sides. DRAW a shape to match each definition.

1. A quadrilateral has 4 sides.
2. A heptagon has 7 sides.
3. A decagon has 10 sides.
4. A parallelogram is a quadrilateral with two pairs of parallel sides.
5. A pentagon has 5 sides.
6. An octagon has 8 sides.
7. A rhombus is a parallelogram whose sides are all of equal length.
8. A nonagon has 9 sides.
9. A trapezoid is a quadrilateral with only one pair of parallel sides.

Clipboard Check

Heidi has a list of shapes and their descriptions, but some of the descriptions are not correct. CHECK the box next to any description that matches the shape.

HINT: A **vertex** is the point where two sides meet.

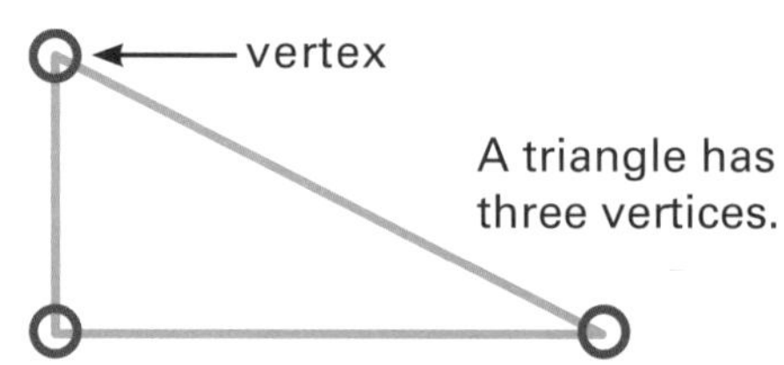

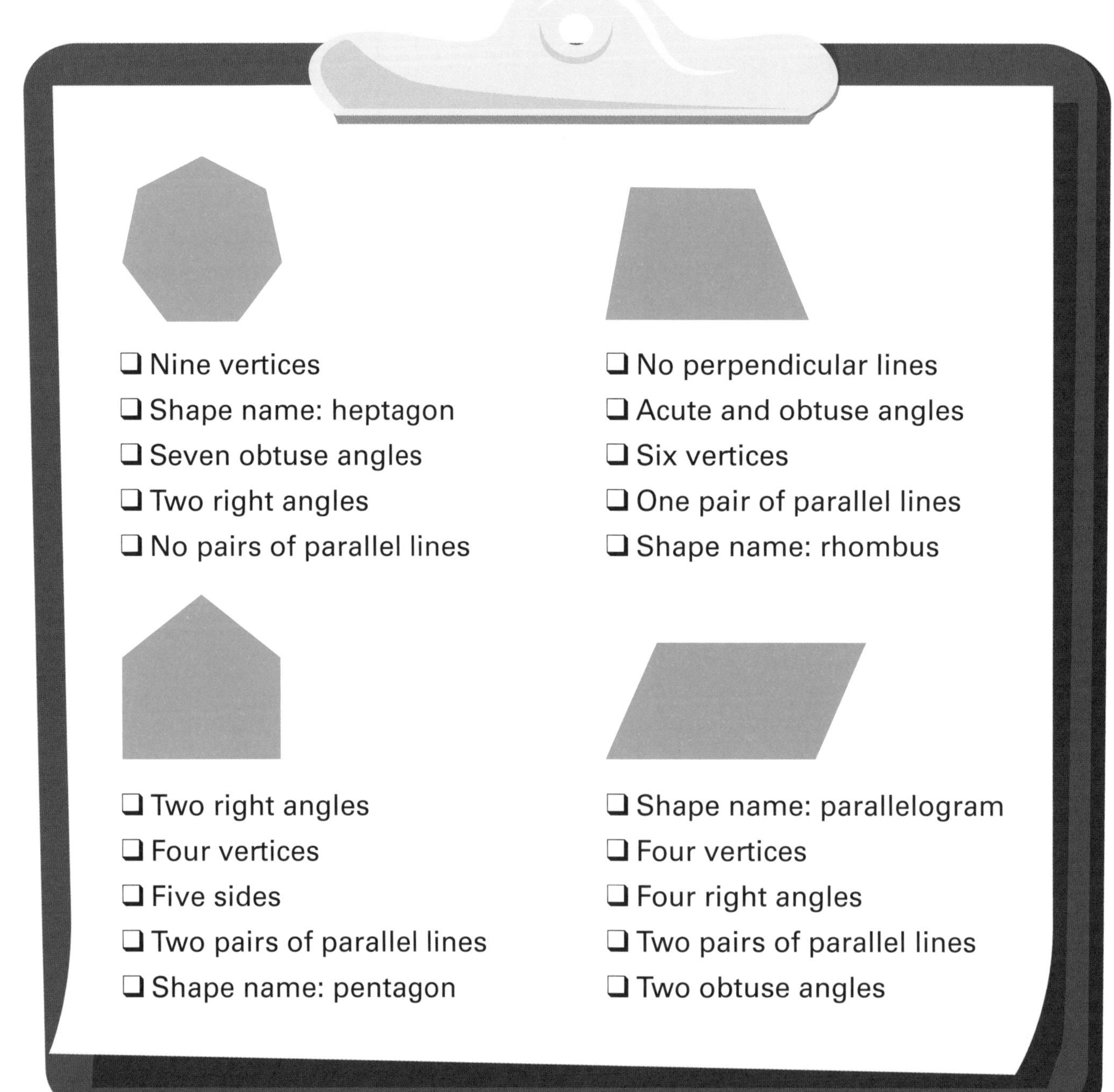

That's My Room!

READ the paragraph, and ANSWER the questions.

HINT: To find the perimeter of something, add the length of each side.

Alyssa's family has just bought a new house. Alyssa knows that her parents will get the biggest room, so she wants to find the next biggest room to claim as her room. There are four bedrooms. Bedroom 1 is 18 feet by 16 feet. Bedroom 2 is $16\frac{1}{2}$ feet by $15\frac{1}{2}$ feet. Bedroom 3 is 14 feet by $18\frac{1}{2}$ feet, and bedroom 4 is $17\frac{1}{2}$ feet by $15\frac{1}{2}$ feet. Alyssa has measured the perimeter of each room and knows which one she wants.

1. Which bedroom does Alyssa want? ____________________
2. Which bedroom will be for Alyssa's parents? ____________________
3. What is the perimeter of Alyssa's room? ____________________

Amusement Adventures

Amy's Amusement Park is a small park that's starting to get crowds for the first time, so Amy has decided to put fences around each of the rides. She has marked where the fences should go and taken measurements in meters (m). WRITE the perimeter of each ride. Then WRITE the total length of fence Amy will need to buy.

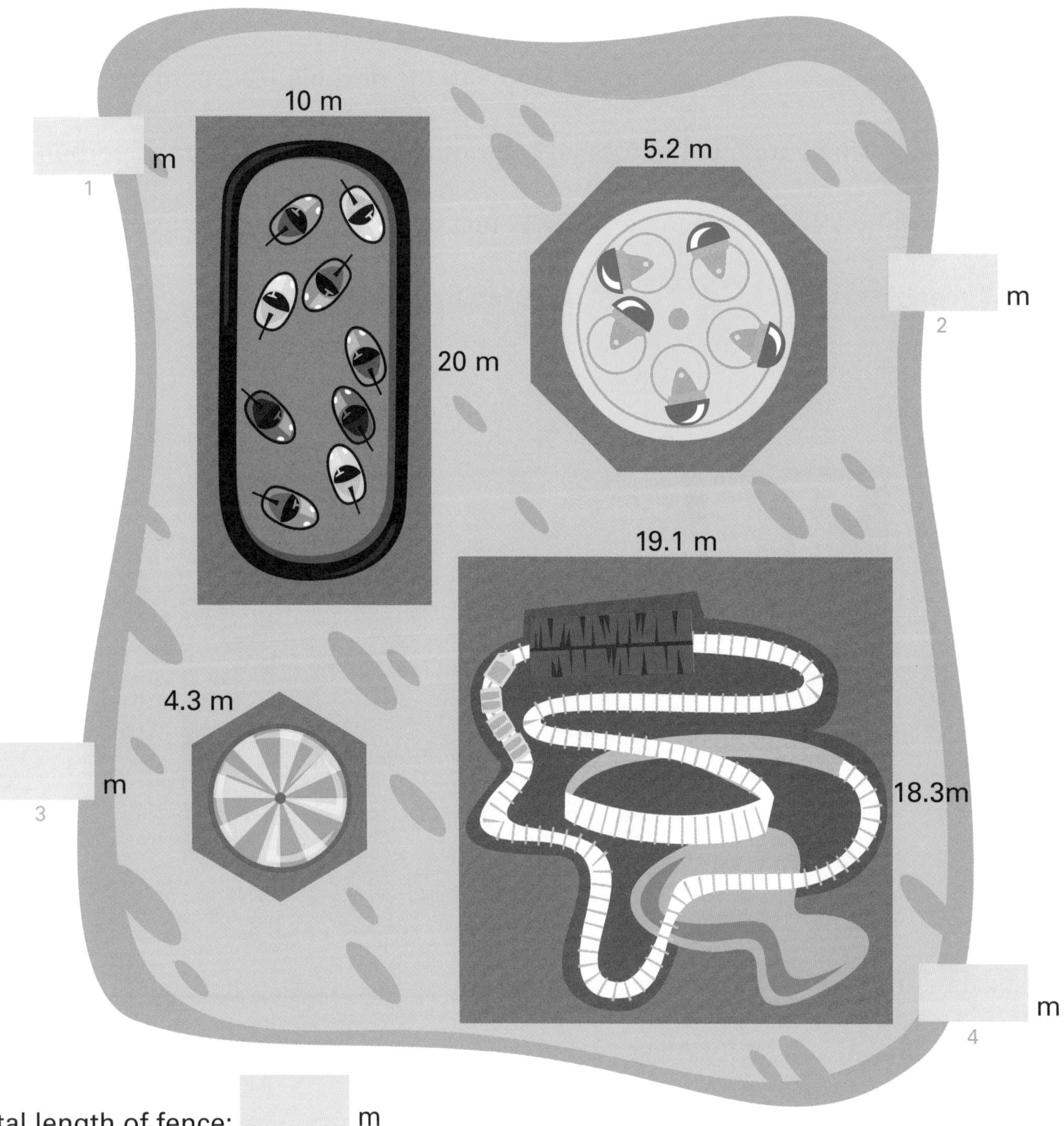

Total length of fence: ______ m
5

Green Acres

The area of one acre is equal to 43,560 square feet or 4,840 square yards. CIRCLE the piece of land that is not one acre.

HINT: Find the area of a rectangle by multiplying the length by the width.

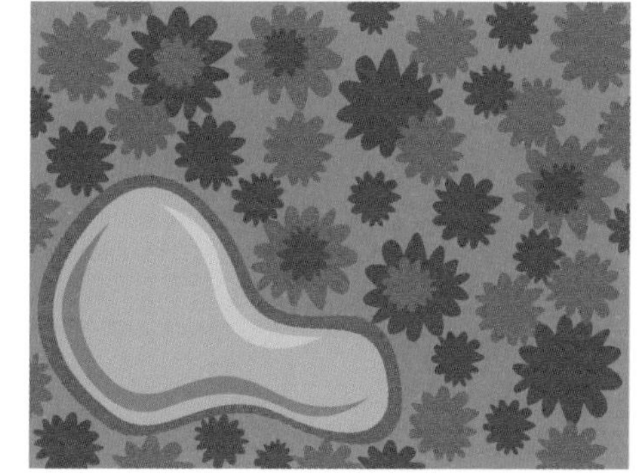

$181\frac{1}{2}$ ft

80 yd

$60\frac{1}{2}$ yd

$75\frac{1}{2}$ yd

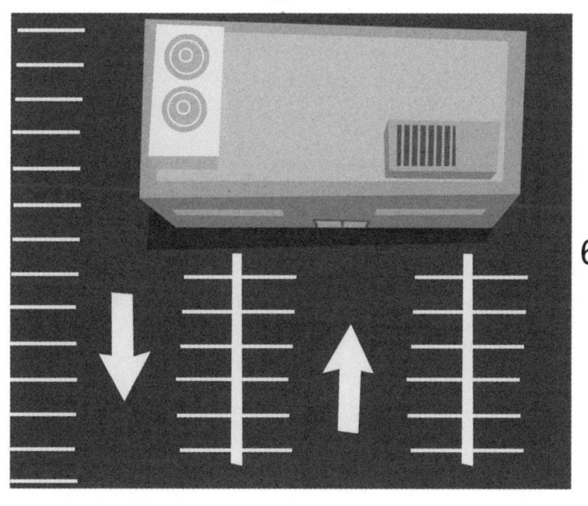

62 yd

$272\frac{1}{4}$ ft

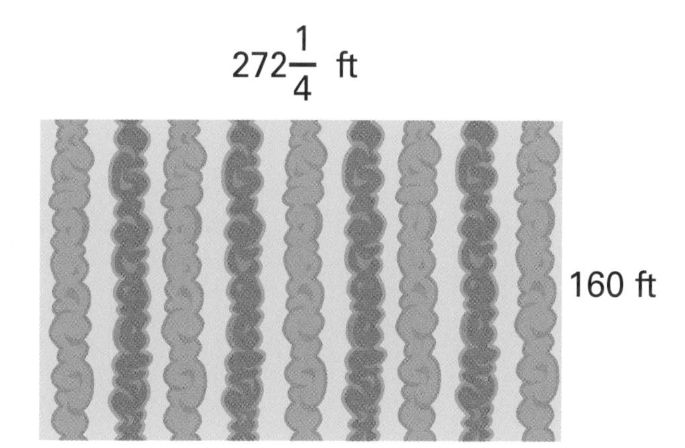

160 ft

Area Carpet

Warren has a most peculiar room, with many angles and corners. He'd like to carpet the room, but he must first figure out the area of the room. He divided the room into different shapes to be able to measure the area. WRITE the area in square meters.

HINT: Find the area of a parallelogram by multiplying the base times the height. Find the area of a triangle by multiplying $\frac{1}{2}$ times the base times the height.

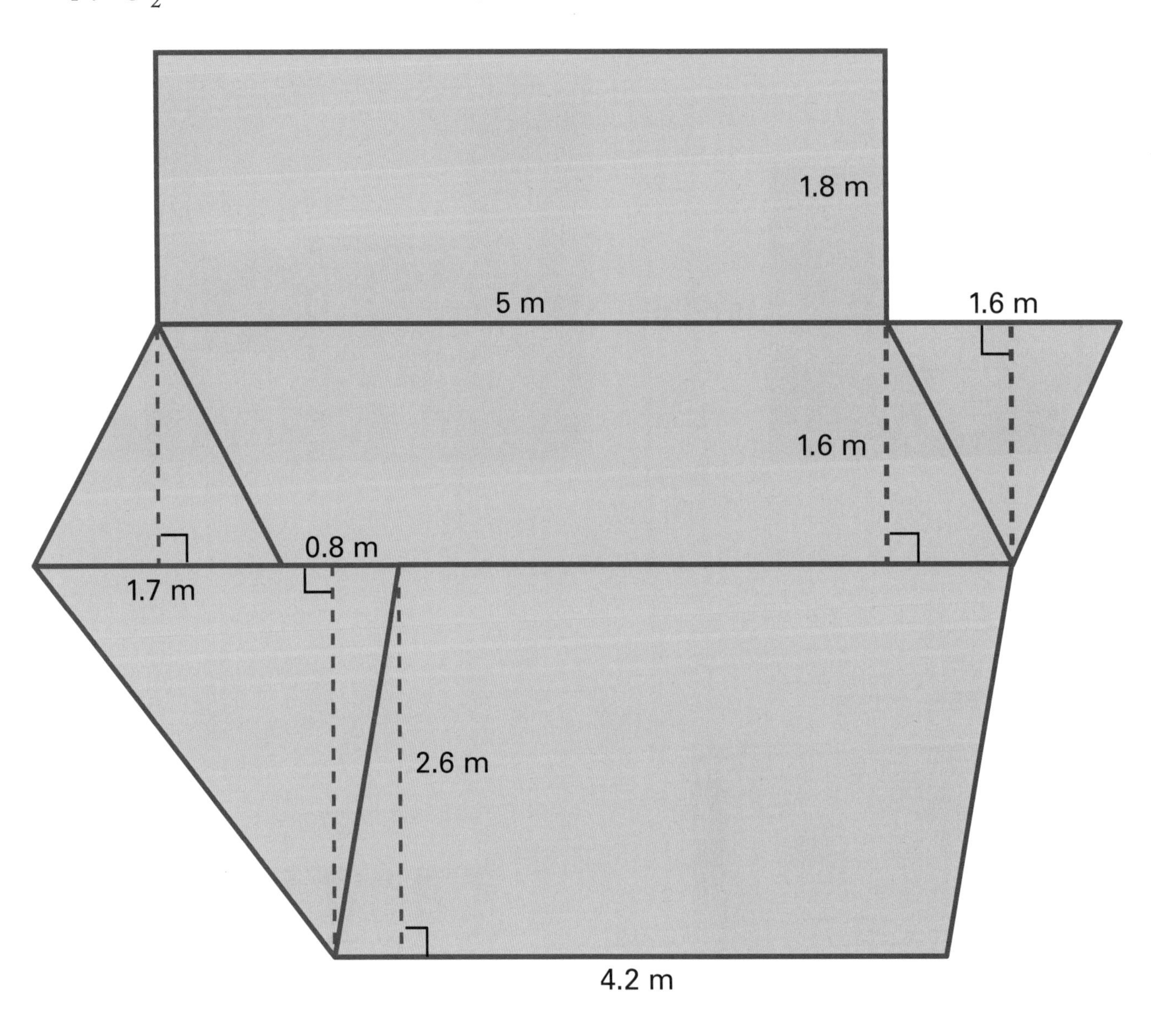

_______ square meters

Box of Chocolates

WRITE the flavor of each piece of chocolate.

Clipboard Check

Tim has a list of shapes and their descriptions, but some of the descriptions are not correct. CHECK the box next to any description that matches the shape.

HINT: In a three-dimensional shape, a vertex is where three or more edges meet. An edge is where two sides meet. A face is the shape formed by the edges.

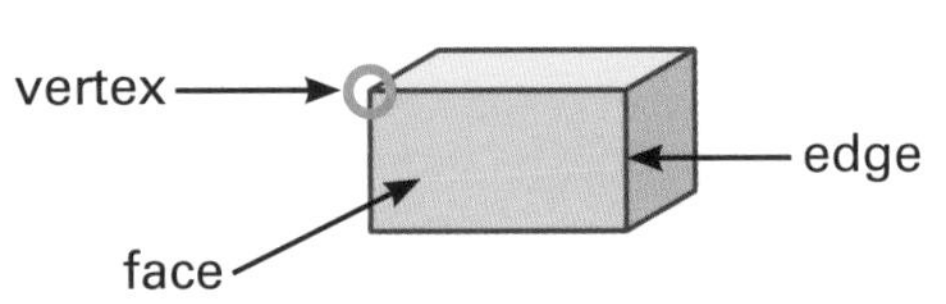

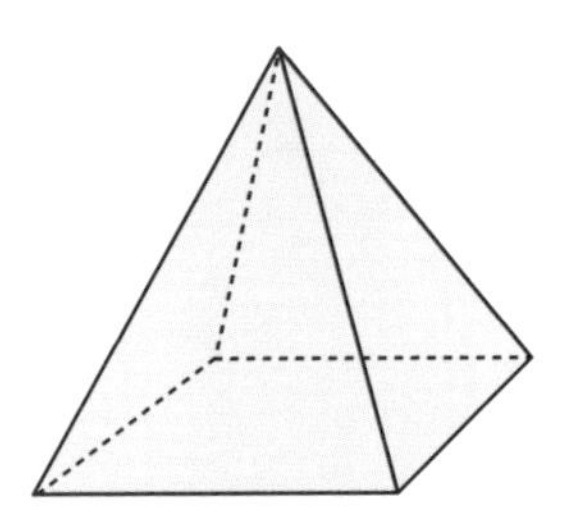

- ❑ Shape name: square pyramid
- ❑ Five vertices
- ❑ Six edges
- ❑ One square face
- ❑ All faces that are triangles

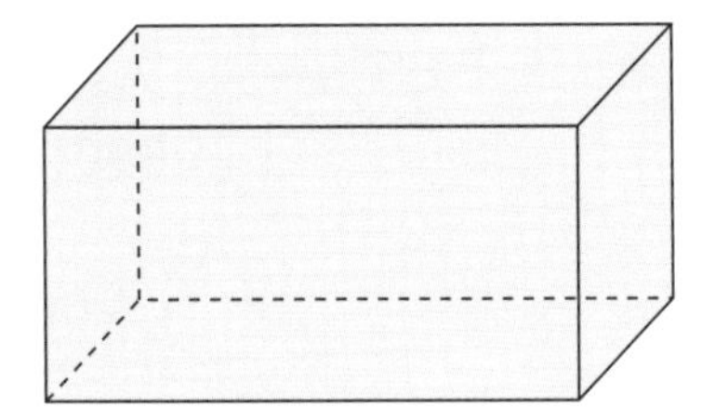

- ❑ Eight edges
- ❑ Shape name: rectangular prism
- ❑ Sides that are parallel
- ❑ Four vertices
- ❑ Six faces

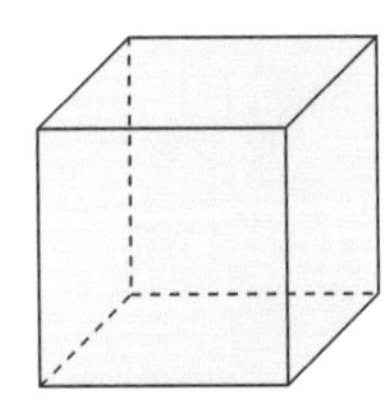

- ❑ Shape name: cube
- ❑ No parallel lines
- ❑ Twelve edges
- ❑ Eight vertices
- ❑ All square faces

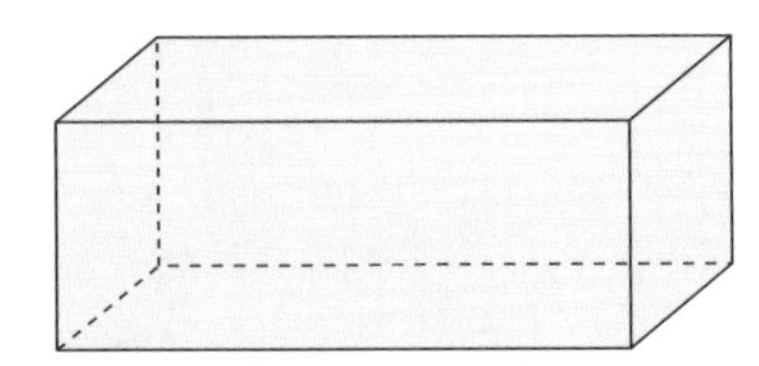

- ❑ Eight faces
- ❑ All faces that are rectangles
- ❑ Shape name: cube
- ❑ Twelve edges
- ❑ Eight vertices

Bug Blocks

The latest craze is building bug houses out of bug blocks. Each bug house set has a different volume. **Volume** is the measure of cubic units that fit inside a space. Think of each bug block as a cubic unit. WRITE the volume of each bug house in cubic units.

Example:

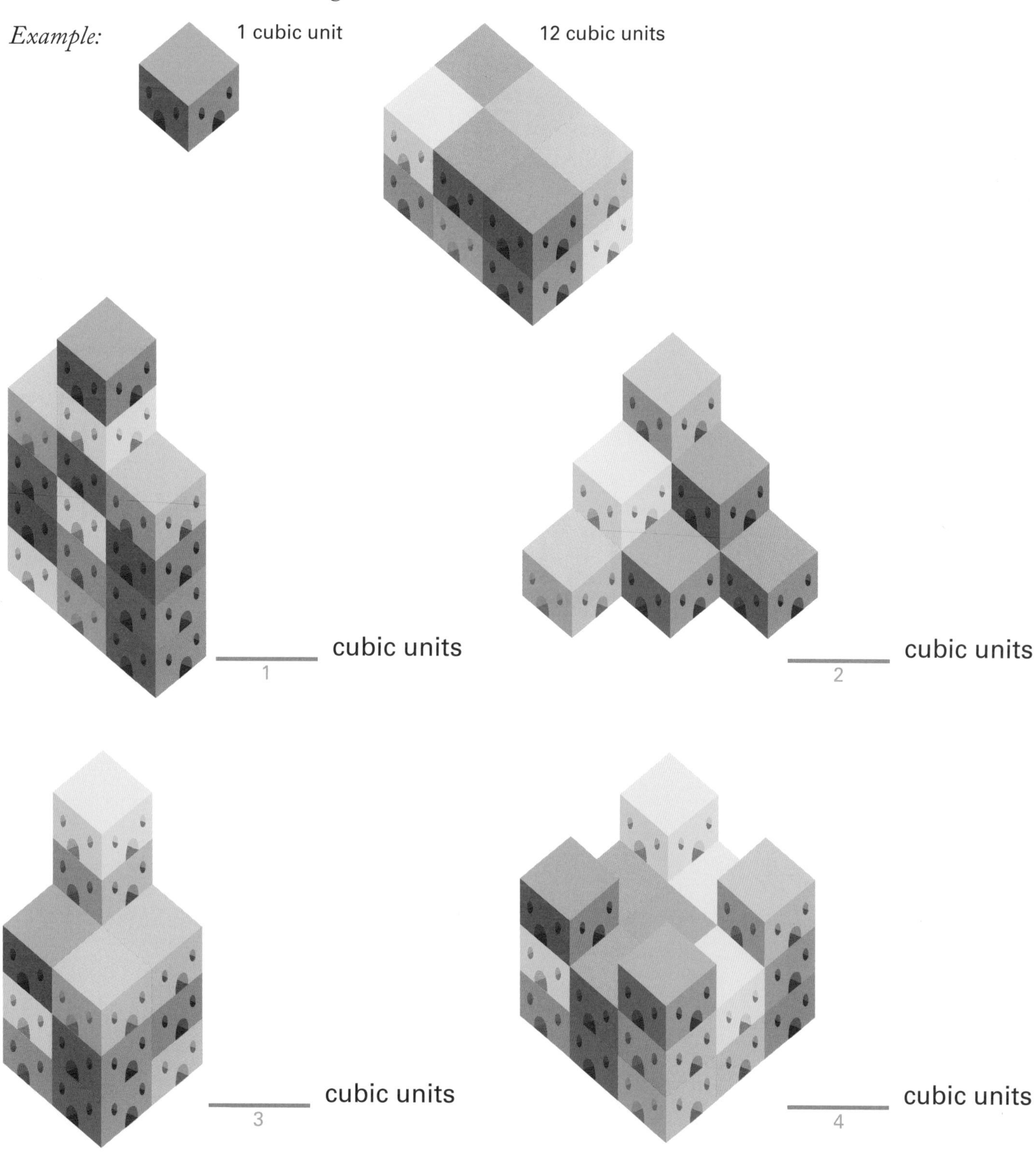

At the Farm

Farmer Brown is thinking about buying a neighborhood farm and is gathering some information about it. WRITE the volume of each building on the farm in cubic feet (ft³).

HINT: Find the volume of each building by multiplying the base times the width times the height.

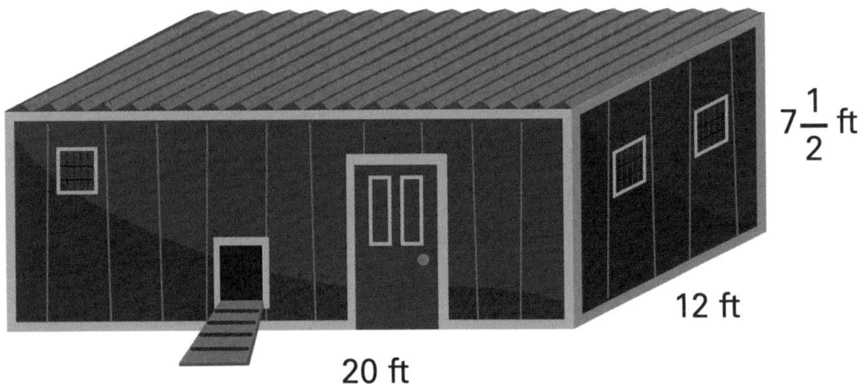

Chicken coop: ________ ft³
1

Greenhouse: ________ ft³
2

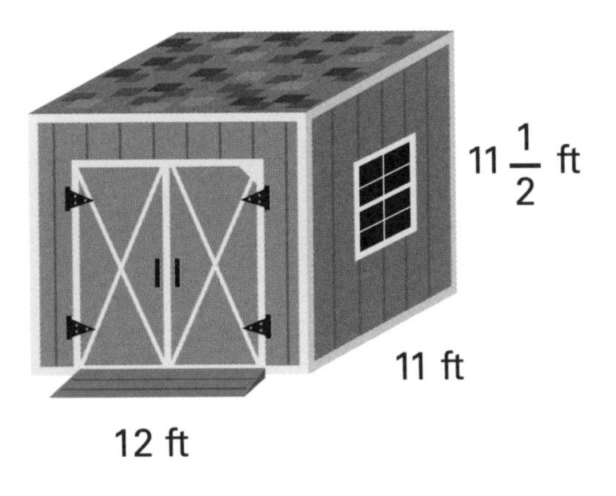

Storage shed: ________ ft³
3

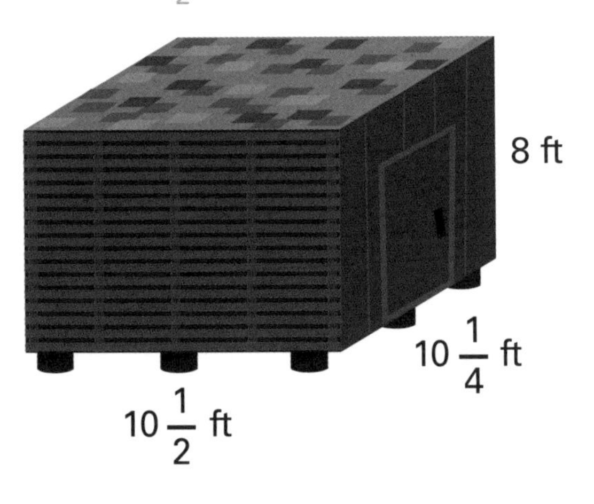

Corn crib: ________ ft³
4

Caged Critters

READ the paragraph, and WRITE the name of each pet next to the correct cage.

Four kids each got a new furry friend and a cage to go along with it. Sara got a pet ferret in a cage with a volume of 8,800 cubic inches. Donnie found a cage with a volume of 1,890 cubic inches for his new pet hamster. Jonathan's new pet parakeet will feel at home in its new cage, which has a volume of 3,528 cubic inches. Yolanda bought a cage with a volume of 4,464 cubic inches for her new pet guinea pig.

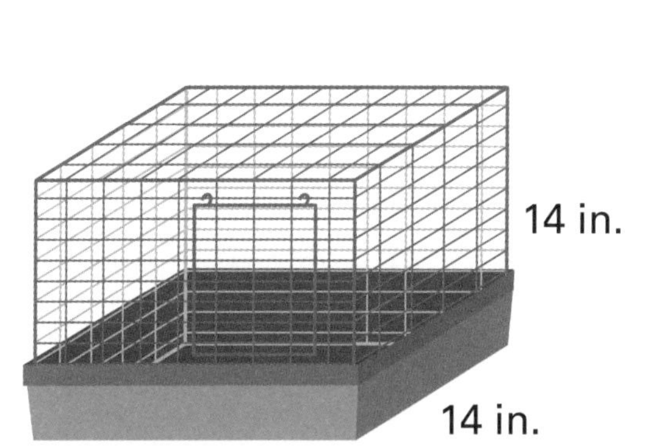

_______________ 1

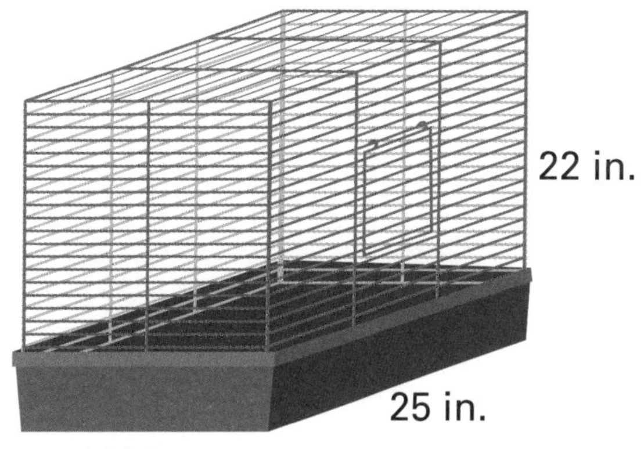

_______________ 2

12 in.
31 in.
12 in.

_______________ 3

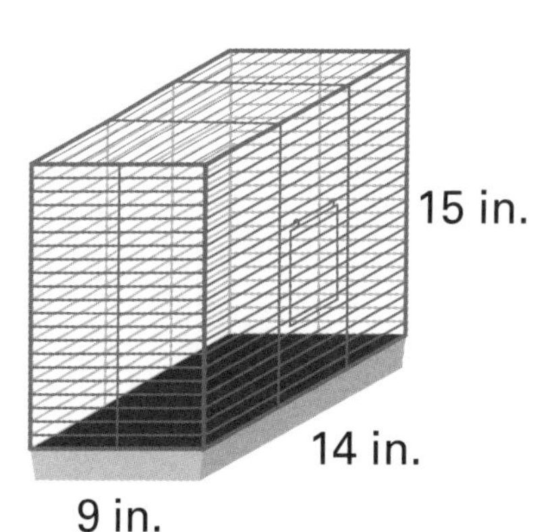

_______________ 4

Shipping Room

READ the paragraph, and WRITE the answer.

A robot needs to be shipped, and you can choose from three different boxes. Box 1 is 10 centimeters by 15 centimeters by 20 centimeters. Box 2 is 10 centimeters by 15 centimeters by 22 centimeters. Box 3 is 12 centimeters by 9 centimeters by 25 centimeters. Write the number of the box that has the right dimensions and volume to fit the robot.

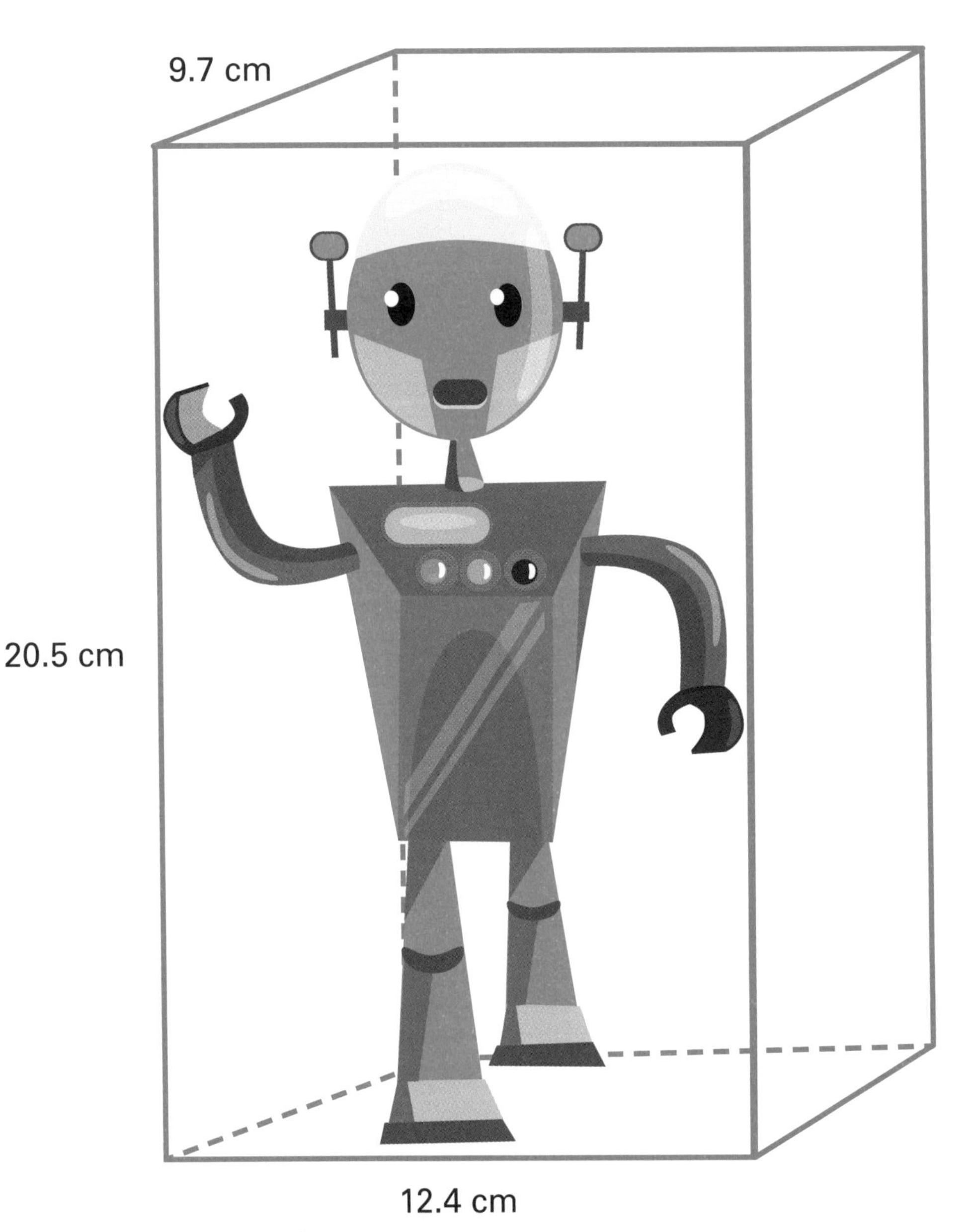

Box ______

What's That Shape?

DRAW each line segment. Then WRITE the name of each shape that you made.

Line segments:

$\overline{AB}$ $\overline{MO}$ $\overline{FG}$ $\overline{BC}$ $\overline{KL}$ $\overline{JN}$ $\overline{CD}$ $\overline{OP}$ $\overline{DE}$

$\overline{GH}$ $\overline{LM}$ $\overline{EF}$ $\overline{NO}$ $\overline{HI}$ $\overline{JK}$ $\overline{LP}$ $\overline{JM}$ $\overline{IA}$

A

I B

H C

G D K

J L

F E

M

____

N P

O

____

Closet Space

Tonya has to clean up all of the things in her room and is hoping she can fit most of her stuff into the closet. To know if she can fit everything, Tonya is taking measurements of her closet. WRITE the perimeter and area of the floor space of the closet. Then WRITE the volume of the closet.

1. Perimeter: ________ ft
2. Area: ________ sq ft
3. Volume: ________ ft^3

Budding Musicians

Kids were asked if they play a musical instrument, and their responses are shown in the line plot. Each X represents one kid. LOOK at the line plot, and ANSWER the questions.

Kids Who Play Instruments

Guitar Trumpet Drums Piano Violin Flute Bass Saxophone

1. How many kids play violin? ________

2. How many kids play drums? ________

3. Exactly four kids play what instrument? ________________

4. What two instruments do the same number of kids play?

____

5. How many more kids play piano than flute? ________

6. How many more kids play violin than bass? ________

7. How many kids play trumpet and saxophone combined? ________

8. If each kid plays only one instrument, how many kids are shown in this graph? ________

Living Situations

Two different groups of kids were asked how many people live in their house. Using the definitions for the line plot on this page to help you, ANSWER the questions about the line plot on the opposite page.

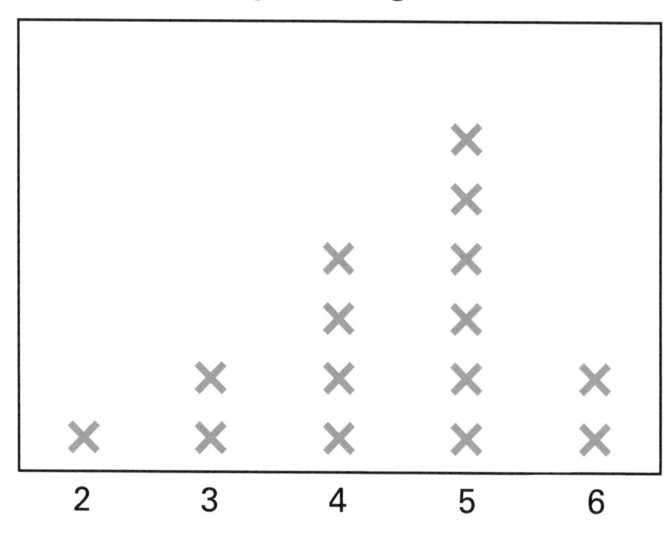

The **range** is the difference between the highest and the lowest numbers of a data set. In this line plot the highest number of people living in the house is 6, and the lowest number is 2. The range of this line plot is 4.

The **median** is the number that is in the middle of a data set. In this line plot, the data set written in order from lowest to highest is 2, 3, 3, 4, 4, 4, 4, 5, 5, 5, 5, 5, 5, 6, 6. The median in this set is 5.

The **mode** is the number or answer that occurs the most. In this line plot, the most kids said that 5 people were living in their house. The mode of this line plot is 5.

The **mean** is the average number in a set of data. It can be found by adding all of the data values together, then dividing by the number of values in the data set. In this line plot, the mean can be found by dividing the total number of people by the number of houses.

First, find the total number of people and houses.

1 house with 2 people = 2 people
2 houses with 3 people = 6 people
4 houses with 4 people = 16 people
6 houses with 5 people = 30 people
2 houses with 6 people = 12 people

There are 15 houses with 66 people. 66 ÷ 15 = 4.4. The mean number of people living in a house is 4.4.

Number of People Living in the House

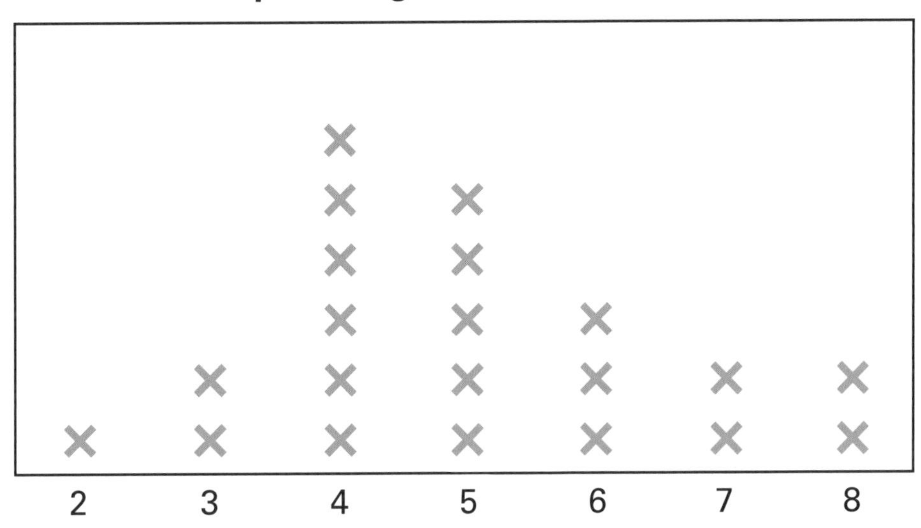

1. The range is ________.

2. The median is ________.

3. The mode is ________.

4. The mean number of people living in the house is ________.

Graph It

ASK 15 people how many books they have read in the past month. RECORD their answers by marking an X on the line plot. Then WRITE the range, median, mode, and mean of your graph.

Books Read in the Past Month

0 1 2 3 4 5 6 7 8 9 10 11 12

The range is ________.

The median is ________.

The mode is ________.

The mean number of books is ________.

Family Membership

The graph shows the number of family memberships to popular city attractions. LOOK at the graph, and ANSWER the questions.

Family Memberships to City Attractions

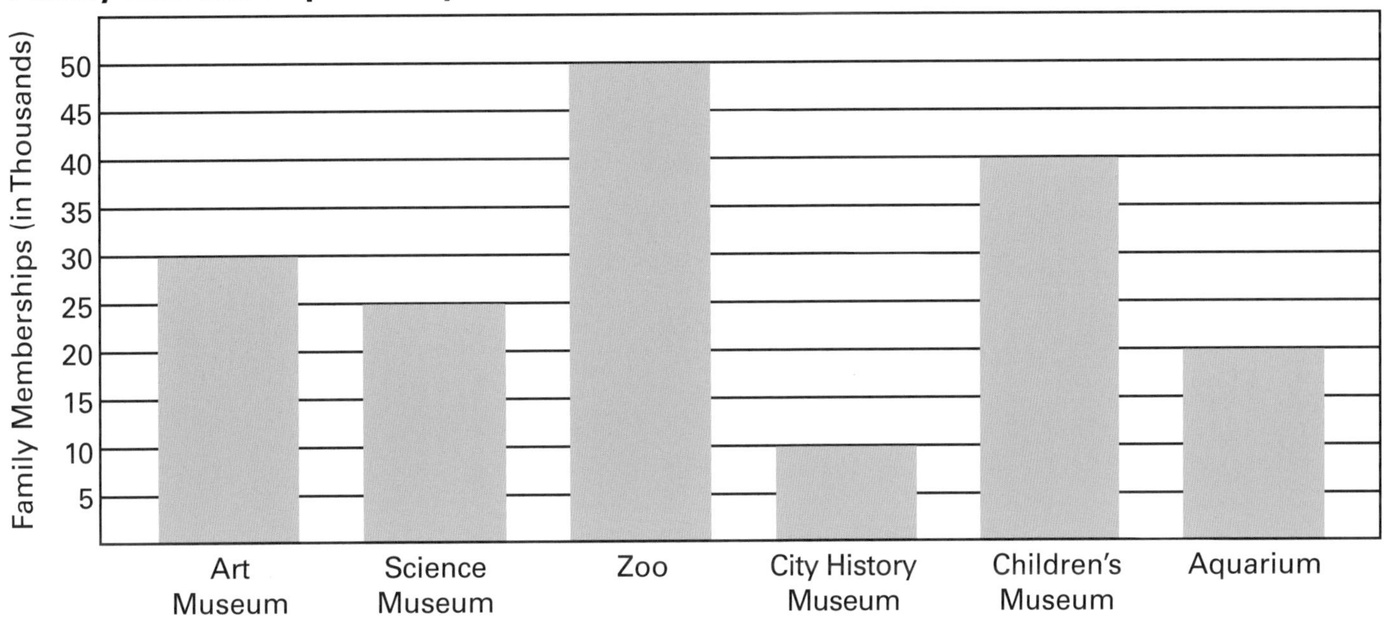

1. How many families have memberships to the science museum? ________

2. How many families have memberships to the children's museum? ________

3. How many more families have memberships to the zoo than to the art museum?

   ____

4. What are the three most popular city attractions for families?

   ____

5. What is the least popular city attraction for families? ____________________

6. How many family memberships have been sold at the four museums combined?

   ____

7. How many family memberships have been sold at the zoo and the aquarium combined?

   ____

8. What is the range in family memberships in this graph? ________

Game Time

This graph shows the number of points scored in the first six games of the season by the two local soccer teams, the Blazers and the Demons. LOOK at the graph, and ANSWER the questions.

Points Scored in Soccer Games

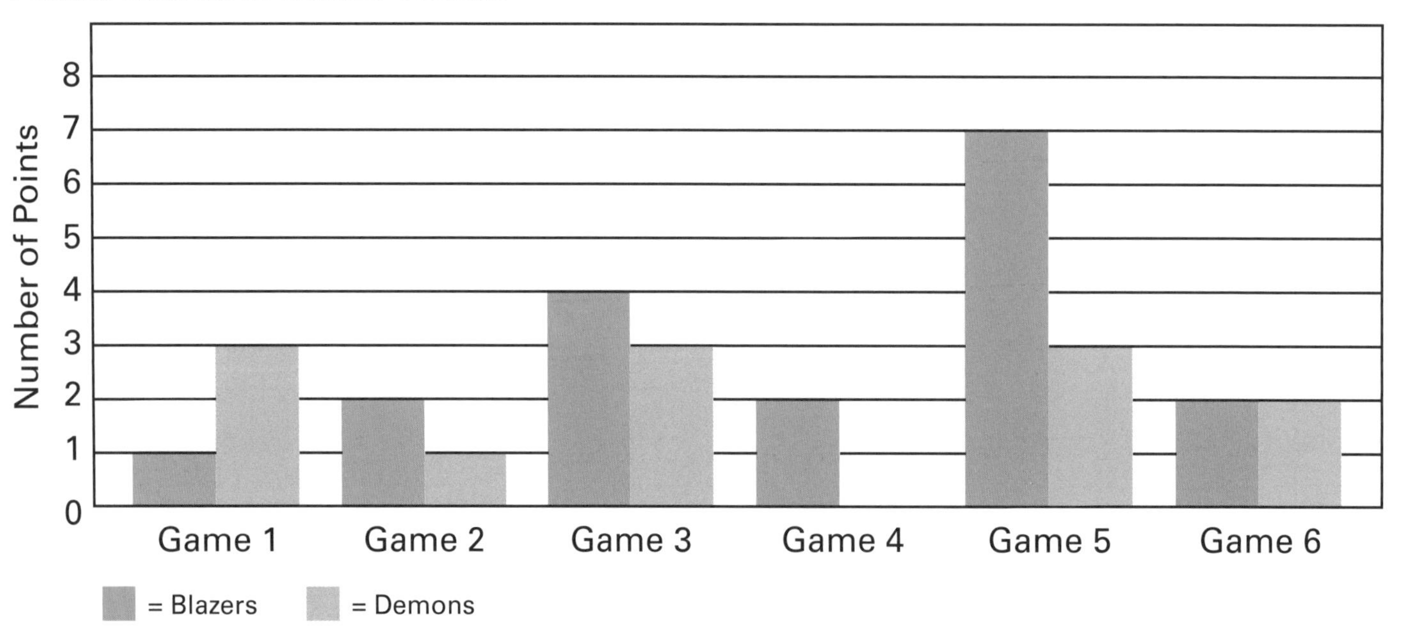

1. How many points did the Blazers score in game 3? ________
2. How many points did the Demons score in game 6? ________
3. In what game did the Blazers score only one point? ____________
4. In what game did the Demons score no points? ____________
5. In one game, the Blazers were playing a team whose best players were all out sick. Judging by the scores in this graph, which game was it? ____________
6. What is the range of scores in this graph? ________
7. What is the mean number of points scored by the Blazers? ________
8. What is the mean number of points scored by the Demons? ________

The Votes Are In

Three candidates ran for mayor. This graph shows the number of votes each got from men and women voters. LOOK at the graph, and ANSWER the questions.

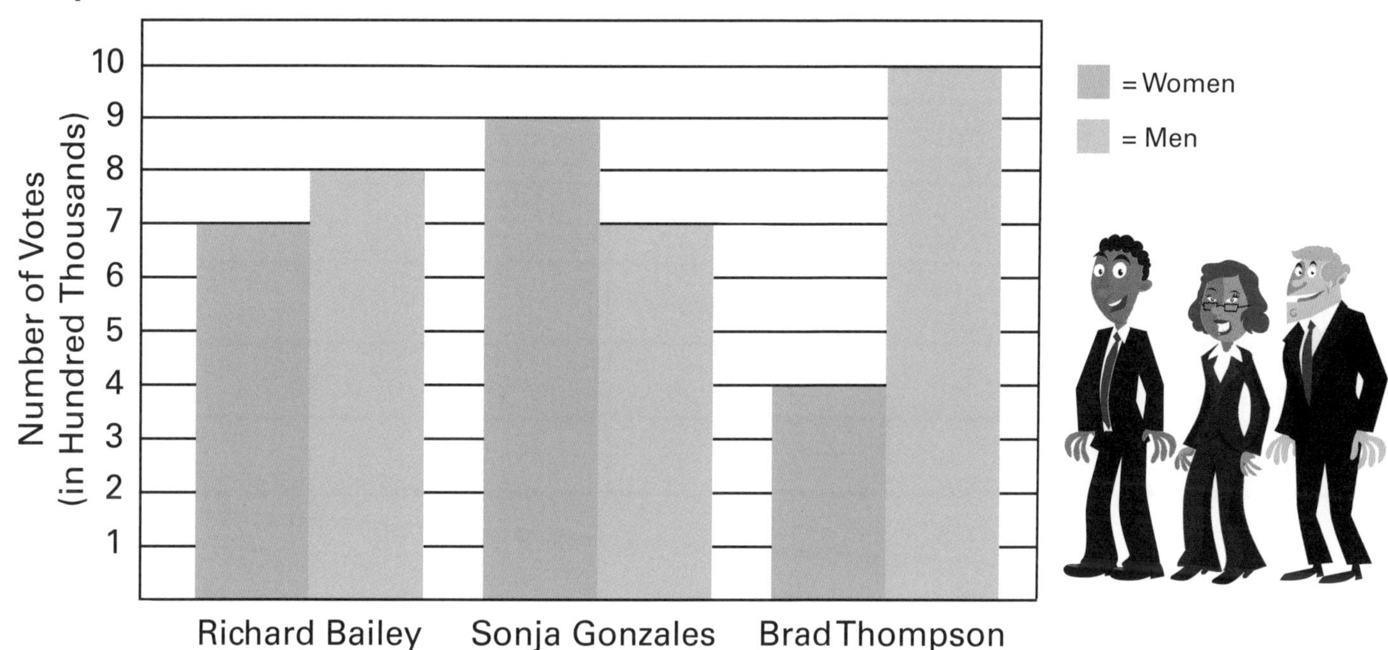

1. How many votes did Richard Bailey get from women voters? ___________
2. How many votes did Brad Thompson get from men voters? ___________
3. Which candidate was more popular among women voters than men voters?

   ____

4. Who got 1,400,000 total votes? ___________________________
5. How many total votes did Richard Bailey get? ___________
6. What is the range of votes from women voters? ___________
7. What is the range of votes from men voters? ___________
8. Who won the election? ___________________________

Graph It

ASK 10 kids and 10 adults about the type of shows they watch in a week. RECORD their answers with tally marks in the chart, one tally mark for each type of show a person watches. Then DRAW the graph.

	Kids	Adults
Cartoons		
News		
Sitcoms		
Dramas		
Reality		

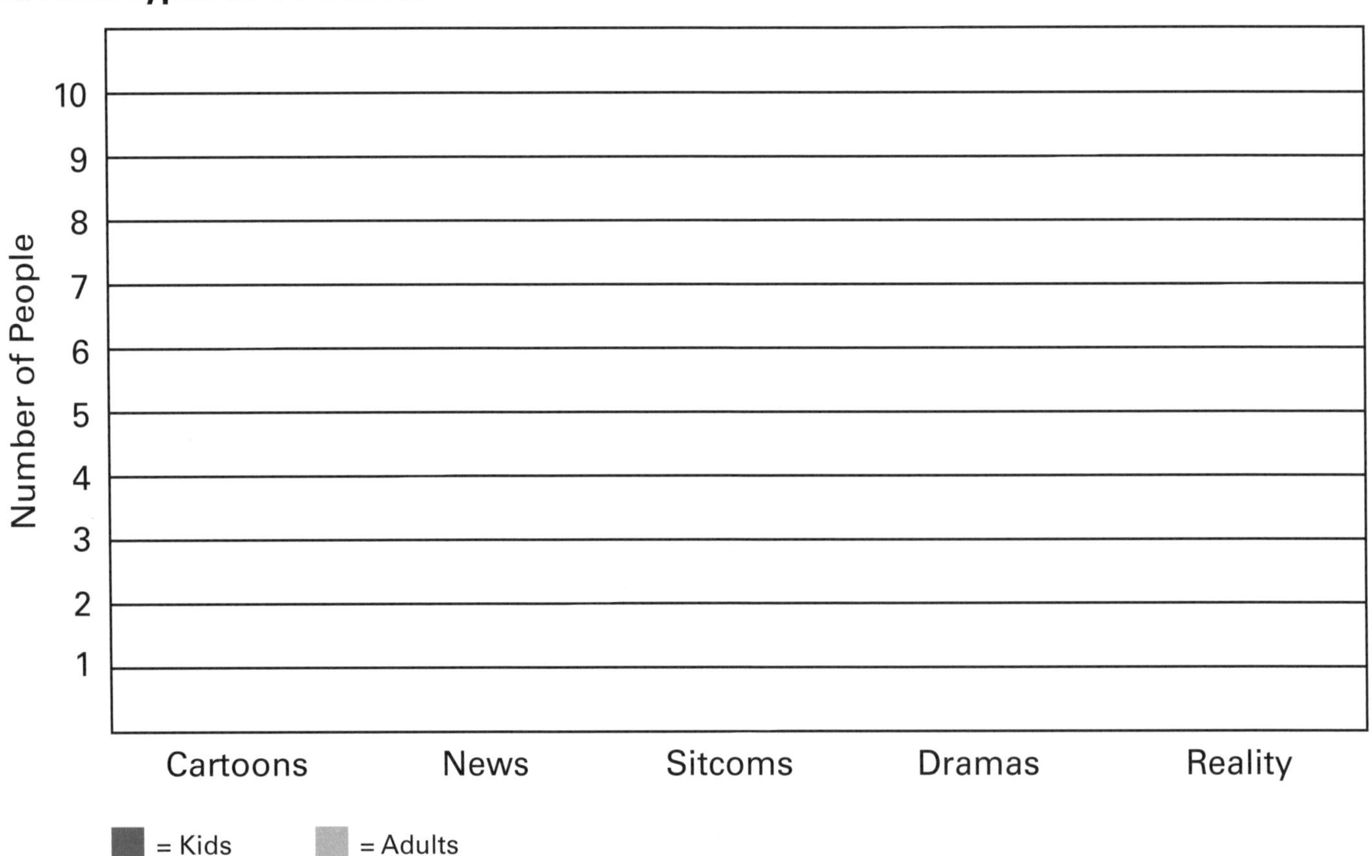

After-School Sports

The two graphs show the number of sports played by boys and girls after school. LOOK at the graphs, and ANSWER the questions.

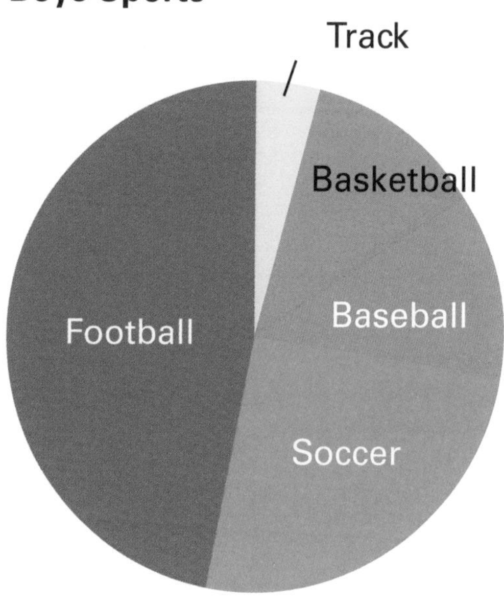

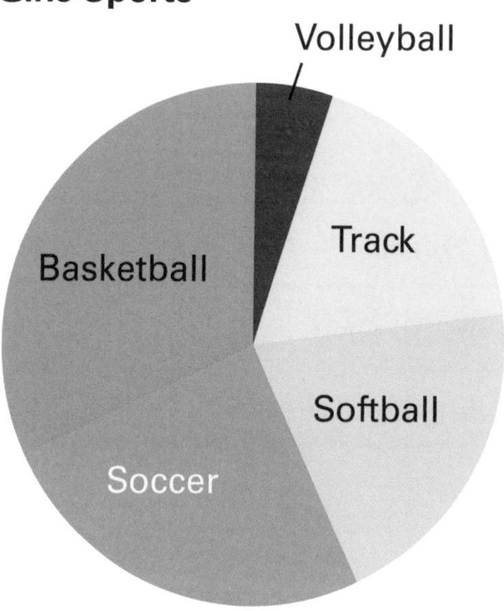

1. About what percentage of boys play football? ______
2. About what percentage of girls play soccer? ______
3. What sport is played by about $\frac{1}{3}$ of girls? ____________
4. What sport is played by about $\frac{1}{25}$ of boys? ____________
5. Which sport do nearly an equal number of girls and boys play? ____________
6. What sport do about 15% more girls play than boys? ____________
7. How likely is it that a boy joining an after-school team will join the track team?

   impossible unlikely likely certain

8. How likely is it that a girl joining an after-school team will join the girl's football team?

   impossible unlikely likely certain

Viewing Audience

Television ratings show the percentage of people from different age groups who watched a particular show. The chart shows the ratings for Thursday evening shows. LOOK at the chart, and WRITE the age-group title for each graph.

Show	Ages 9–17	Ages 18–34	Ages 35–49
Family Flies	14%	7%	3%
The B Team	24%	23%	11%
King of the Grill	5%	8%	21%
The Clinging Detective	10%	17%	42%
Miami Mice	46%	10%	4%
Twin Creeks	1%	35%	19%

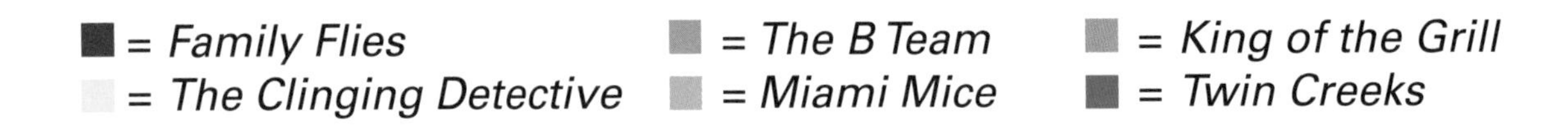

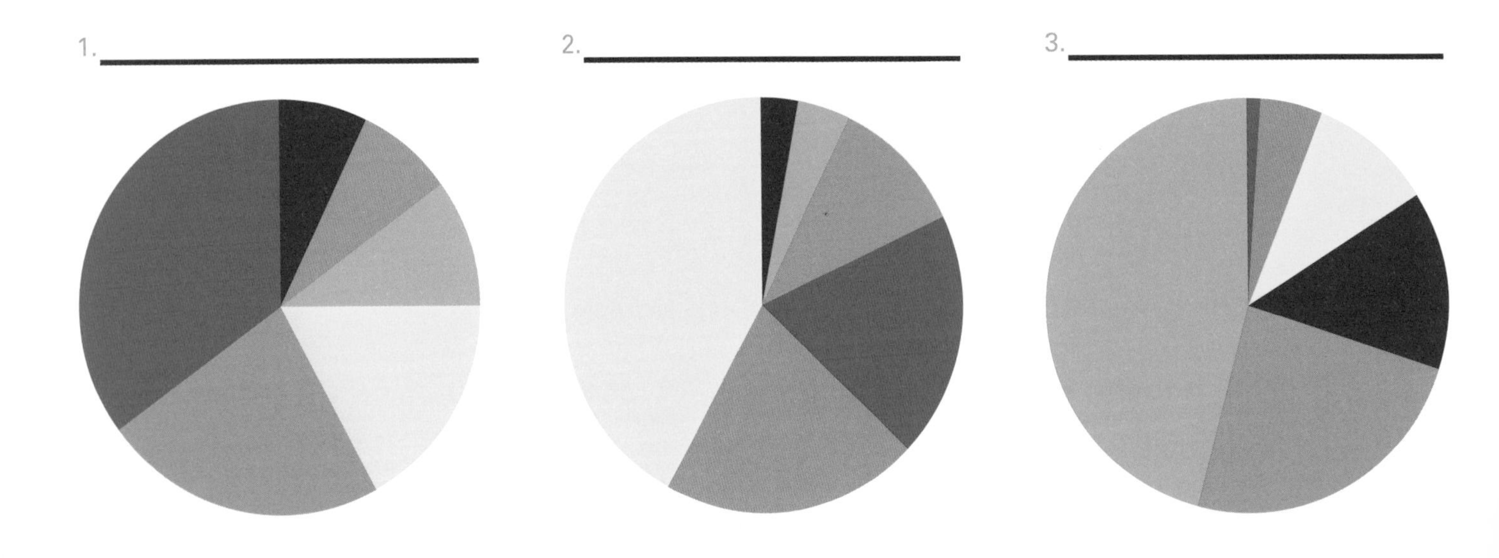

Sandwich Shop

The owners of Sal's Sandwich Shop are changing the menu but want to keep the most popular sandwiches. They graphed the sandwich purchases of 500 customers. LOOK at the graph, and ANSWER the questions.

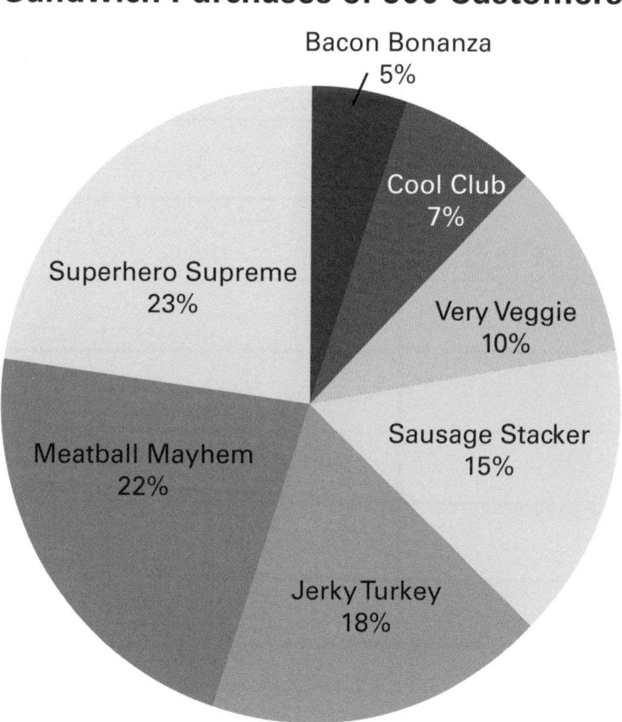

How many people bought each sandwich?

1. Superhero Supreme ______
2. Sausage Stacker ______
3. Meatball Mayhem ______
4. Very Veggie ______
5. Cool Club ______
6. Bacon Bonanza ______
7. Jerky Turkey ______

8. What are the two most popular sandwiches? ______________________

____

9. Based on this graph, what two sandwiches should the owners of Sal's Sandwich Shop replace with new sandwiches? ______________________

Graph It

ASK 20 people their favorite color. RECORD their answers with tally marks in the chart. Then COLOR the graph according to the favorite colors, and WRITE the percentage of people that like each color.

Color	Number of People	Percent of People

Favorite Color

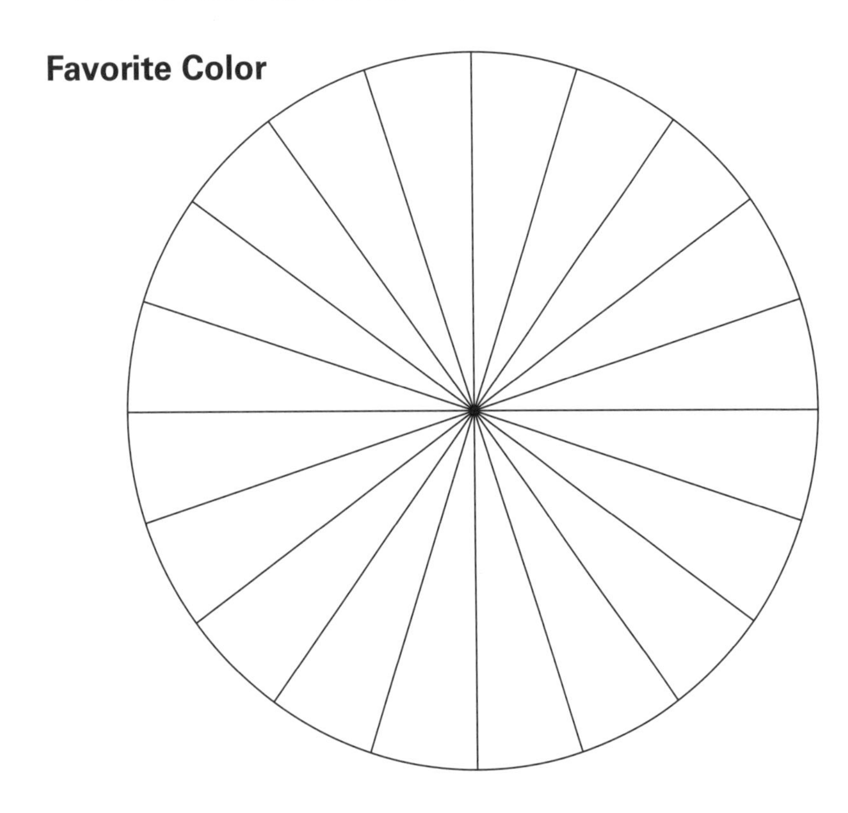

What's in a Name?

This graph shows the popularity of a name as it changed over time. The names are ranked in popularity from 1 to 100. LOOK at the graph, and ANSWER the questions.

HINT: A name only appears on this graph in the years that it was in the top 100 names.

Name Rank by Year

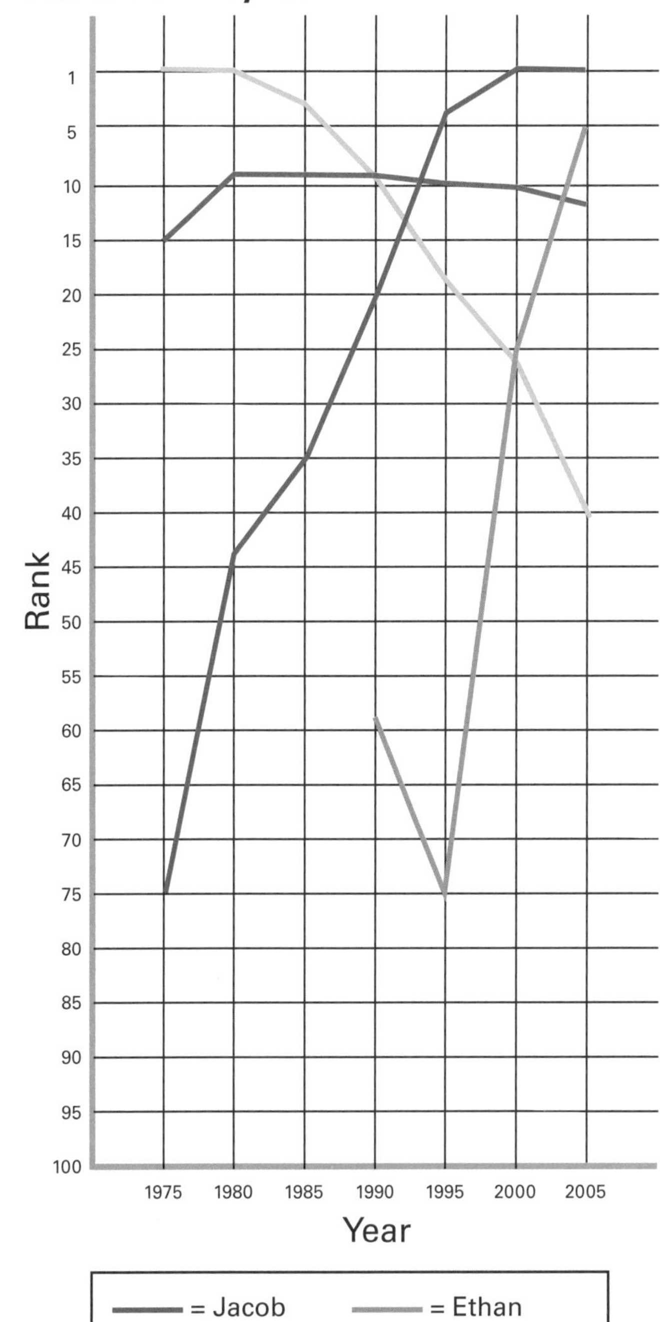

1. What name was ranked number 35 in 1985? ____________

2. What name was ranked number 5 in 2005? ____________

3. What two names have been ranked at number 1?

____

4. What two names have been ranked at number 75?

____

5. What name did not rank in the top 100 until 1990? ____________

6. What name has had a rank that remained about the same over this 30-year period? ____________

7. In what year were Jennifer and Elizabeth similarly ranked? ____________

8. In what year were Jennifer and Ethan similarly ranked? ____________

Population Shift

USE the data on this page to DRAW the graph on the opposite page.

HINT: Round the populations to the nearest thousand, and estimate the placement of each point on the graph.

	Miami, FL	St. Louis, MO	Pittsburgh, PA
1930	110,637	821,960	669,817
1940	172,172	816,048	671,659
1950	249,276	856,796	676,806
1960	291,688	750,026	604,332
1970	334,859	622,236	520,117
1980	346,865	453,085	423,938
1990	358,548	396,685	369,879
2000	362,470	348,189	334,563

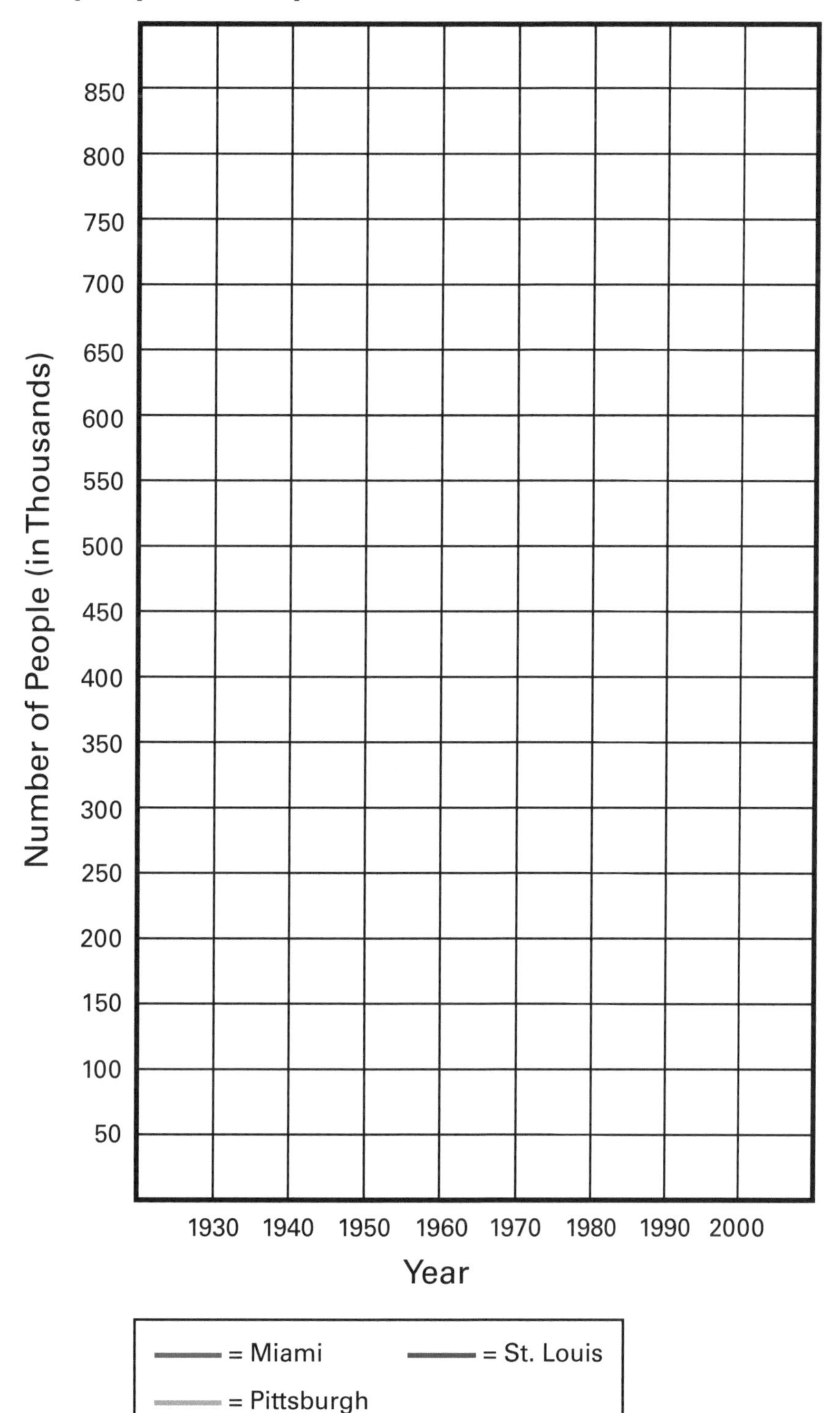
City Population by Year
Number of People (in Thousands)
850
800
750
700
650
600
550
500
450
400
350
300
250
200
150
100
50
1930
1940
1950
1960
1970
1980
1990
2000
Year
= Miami
= St. Louis
= Pittsburgh

On the Phone

This graph shows the number of minutes spent on a cell phone over a six-month period. LOOK at the graph, and ANSWER the questions.

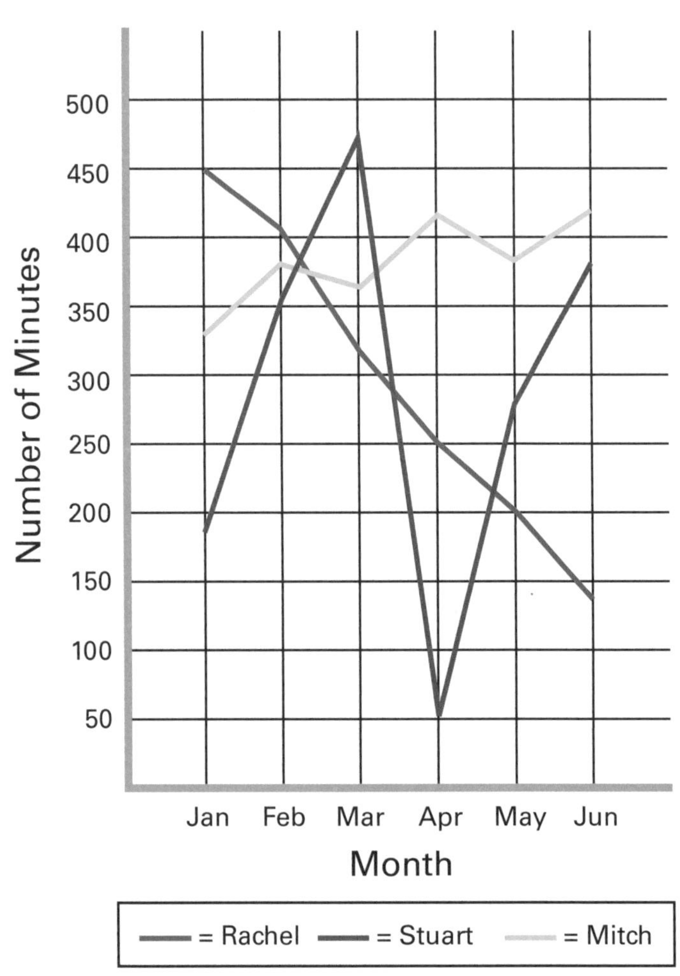

1. About how many minutes did Stuart use in February? ____________
2. About how many minutes did Rachel use in May? ____________
3. Who used the most minutes in April? ____________
4. Whose cell phone use is pretty consistent from month to month?

   ____
5. Stuart lost his phone for most of a month but then found it again. Which month was it? ____________
6. How likely is it that Rachel will use her cell phone for around 400 minutes in July?

   impossible unlikely likely certain
7. How likely is it that Mitch will use his cell phone for around 400 minutes in July?

   impossible unlikely likely certain
8. Stuart used his phone for 182, 351, 468, 53, 274, and 382 minutes each month over the 6-month period. What is his mean number of minutes? ____________

Graph Grabber

An **apiary** is a place where colonies of honeybees are kept. Which graph works best with the title "Honey Produced per Aviary"? CIRCLE the correct graph.

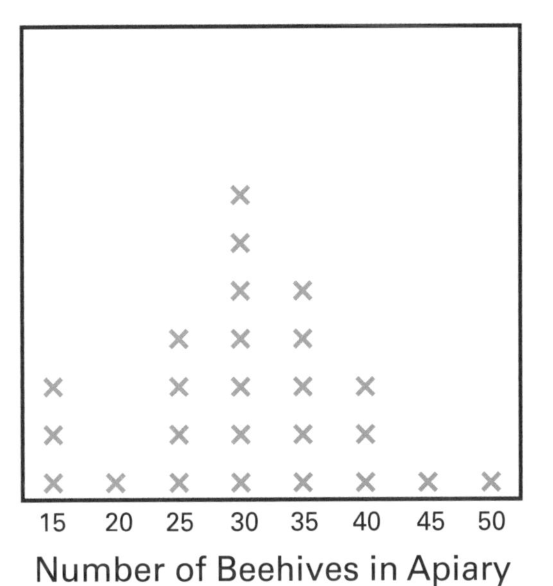

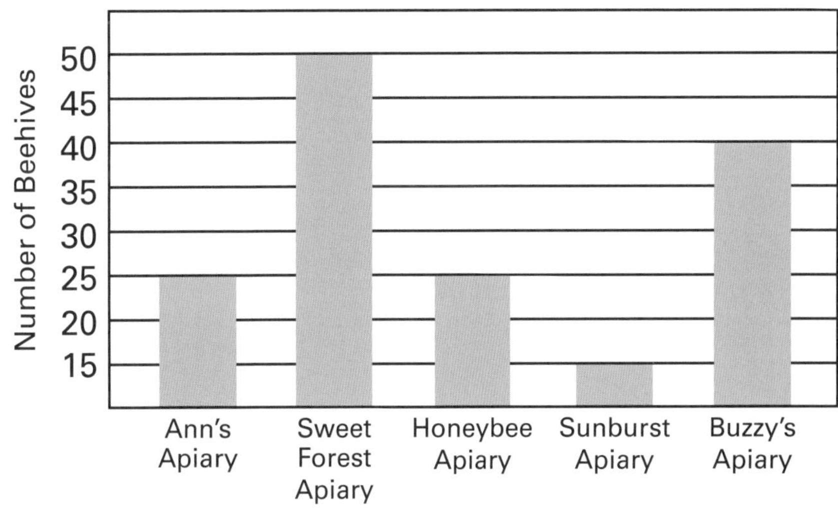

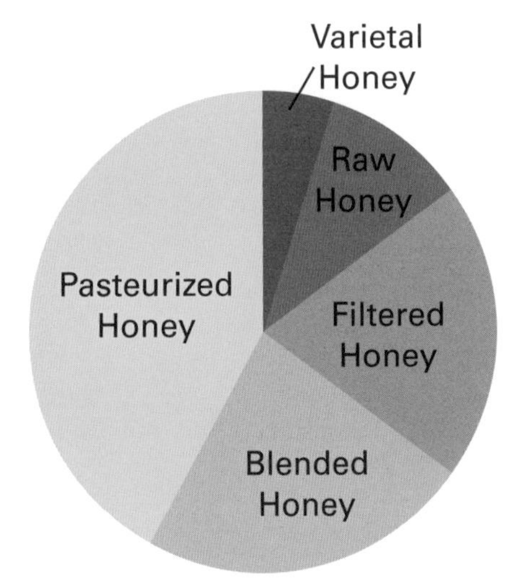

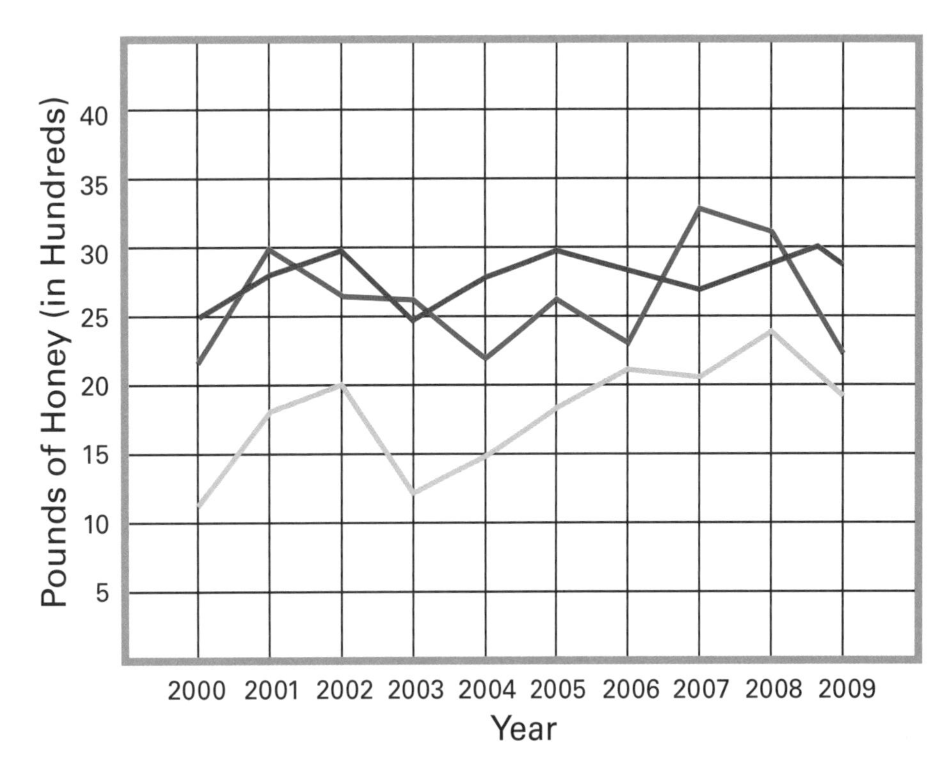

Graph It

Now it's your turn to pick the topic of your graph. Think of a question you're curious about, and ASK 20 people their answer. RECORD their answers with tally marks in the chart. Then DRAW the graph and give it a title.

Answer	Number of People

Title: ____________________

Page 3

Player 1	
800 ×	40,000
70 ×	35,000
4 ×	20,000
Total Score	95,000

Player 2	
1,000 ×	50,000
200 ×	100,000
1 ×	5,000
Total Score	155,000

Player 3	
90 ×	4,500
30 ×	15,000
20 ×	100,000
Total Score	119,500

Player 4	
2,000 ×	100,000
11 ×	5,500
9 ×	45,000
Total Score	150,500

Page 4

1	1	1
× 3	× 4	× 5
3	4	5
× 3	× 4	× 5
9	16	25
× 3	× 4	× 5
27	64	125
× 3	× 4	× 5
81	256	625

Page 5

1. 375
2. 360
3. 252
4. 160
5. 330
6. 480
7. 192
8. 220
9. 480
10. 340
11. 180
12. 190
13. 225
14. 520
15. 312
16. 200

Page 6

1. 77
2. 120
3. 112
4. 144
5. 175
6. 372
7. 322
8. 708
9. 441
10. 888
11. 616
12. 1,164

Page 7

1. 480, 1,260, 432, 585
2. 408, 828, 828, 1,350
3. 600, 900, 450, 1,125
4. 1,248, 1,368, 1,170, 1,890

Page 8

1. 11,835
2. 10,260
3. 19,850
4. 18,090
5. 20,370
6. 17,640
7. 14,760
8. 36,250
9. 14,700
10. 5,600

Page 9

1. 173,784
2. 86,870
3. 171,080
4. 130,500
5. 173,568
6. 302,744
7. 159,064
8. 199,892

Page 10

Page 11

1. 20,000
2. 9,000
3. 42,000
4. 16,000
5. 15,000
6. 49,000
7. 4,000
8. 15,000
9. 36,000
10. 30,000
11. 12,000
12. 16,000

Page 12

404 × 192 → 400 × 200 = 80,000; calculator: 77,568

769 × 329 → 800 × 300 = 240,000; calculator: 153,001

957 × 521 → 1,000 × 500 = 500,000; calculator: 498,597

345 × 466 → 300 × 500 = 150,000; calculator: 125,983

2,879 × 289 → 2,900 × 300 = 870,000; calculator: 1,820,686

1,664 × 743 → 1,700 × 700 = 1,190,000; calculator: 1,236,352

Page 13

1. Kenneth: 8, Courtney: 16, Diana: 28, Aaron: 32
2. Joseph: 10, Alyssa: 15, Isaac: 25, Cassandra: 45
3. Alejandro: 18, Victoria: 42, Hunter: 54, Faith: 66

Page 14

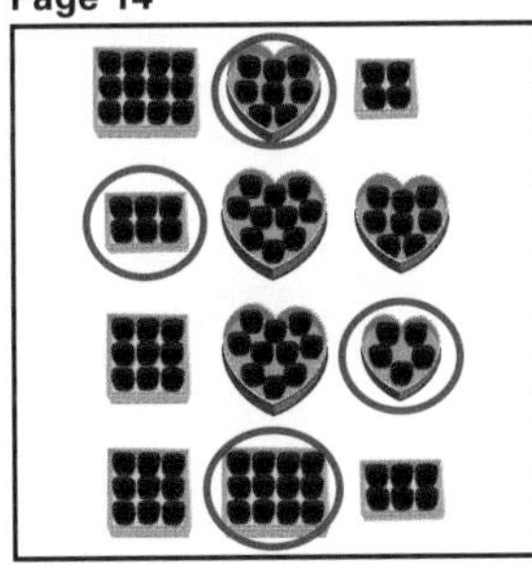

Page 15

1. 532,900, 563,925
2. 874,175, 933,670
3. 692,040, 719,050
4. 764,675, 774,530

Page 16

1. 135,360
2. 164,160
3. 123,840
4. 151,200
5. 112,320
6. 190,080
7. 142,560
8. 174,240

Page 17

1. 60, 50, 40
2. 300, 200, 100
3. 250, 200, 150
4. 4,000, 3,500, 2,000

Page 18

324	512	3,750
÷ 3	÷ 4	÷ 5
108	128	750
÷ 3	÷ 4	÷ 5
36	32	150
÷ 3	÷ 4	÷ 5
12	8	30
÷ 3	÷ 4	÷ 5
4	2	6

Page 19

1. 63
2. 196
3. 223
4. 631

Page 20

1. 9,435
2. 8,128
3. 11,592
4. 10,663
5. 9,008
6. 7,950
7. 13,624
8. 10,708
9. 8,712
10. 12,558

Page 21

1. 7
2. 6
3. 9
4. 11
5. 13
6. 8
7. 5
8. 18
9. 4
10. 9
11. 12
12. 14

Page 22

1. 17, 5
2. 22, 17
3. 15, 3
4. 14, 11
5. 13, 50
6. 9, 45

Page 23

1. 47, 4
2. 28, 16
3. 37, 5
4. 46, 10
5. 160, 5
6. 497, 7

Page 24

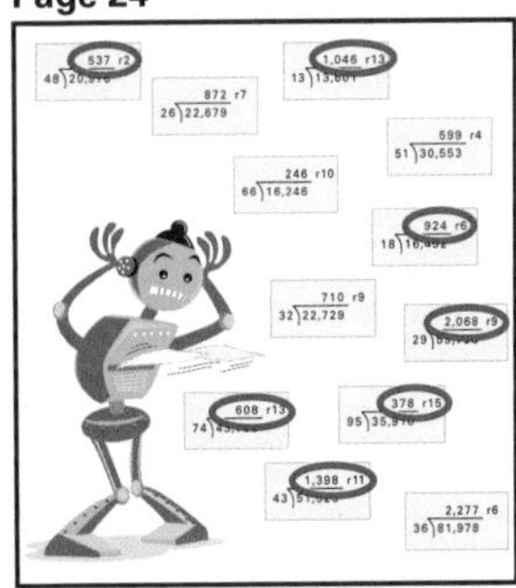

Page 25

1. 3,000
2. 2,000
3. 9,000
4. 5,000
5. 4,000
6. 35,000

Page 26

9)36,423 → 9)36,000 = 4,000; calculator: 4,047

15)44,835 → 15)45,000 = 3,000; calculator: 1,589

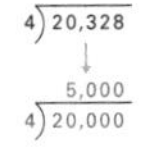

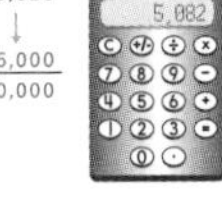

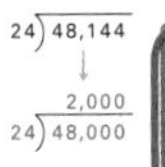

70)69,580 → 70)70,000 = 1,000

Page 27

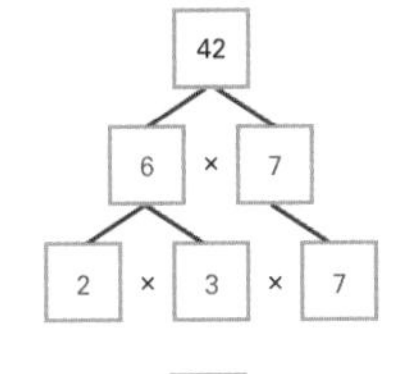

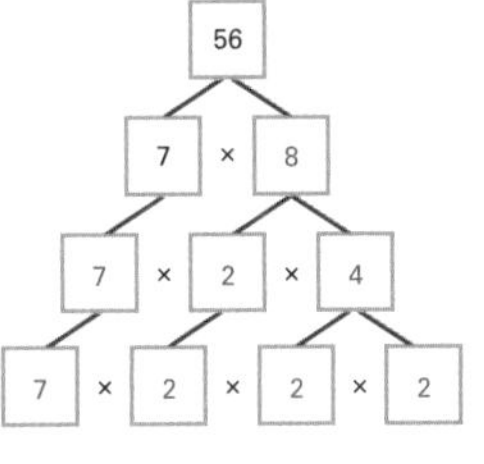

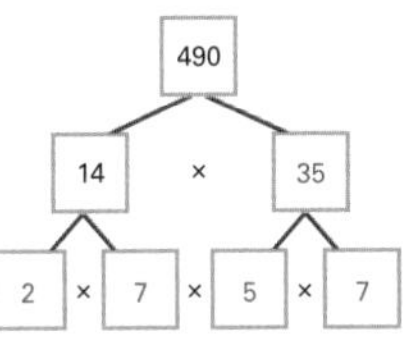

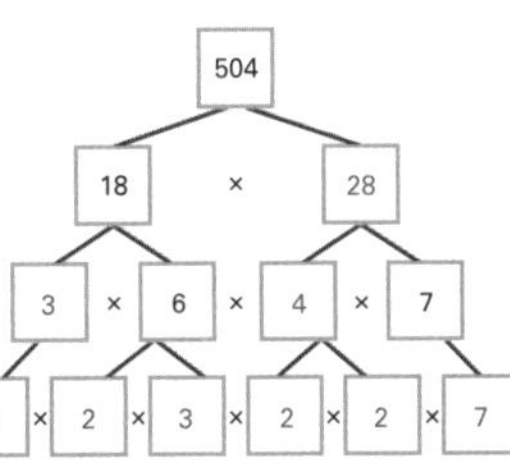

Page 28

1	**2**	**3**	4	**5**	6	**7**	8	9	10
11	12	**13**	14	15	16	**17**	18	**19**	20
21	22	**23**	24	25	26	27	28	**29**	30
31	32	33	34	35	36	**37**	38	39	40
41	42	**43**	44	45	46	**47**	48	49	50
51	52	**53**	54	55	56	57	58	**59**	60
61	62	63	64	65	66	**67**	68	69	70
71	72	**73**	74	75	76	77	78	**79**	80
81	82	**83**	84	85	86	87	88	**89**	90
91	92	93	94	95	96	**97**	98	99	100

Answers

Page 29
1. 513, 5, 13
2. 639, 6, 39
3. 826, 8, 26
4. 908, 9, 8

Page 30
1. 219
2. 25
3. 46
4. 10

Page 31

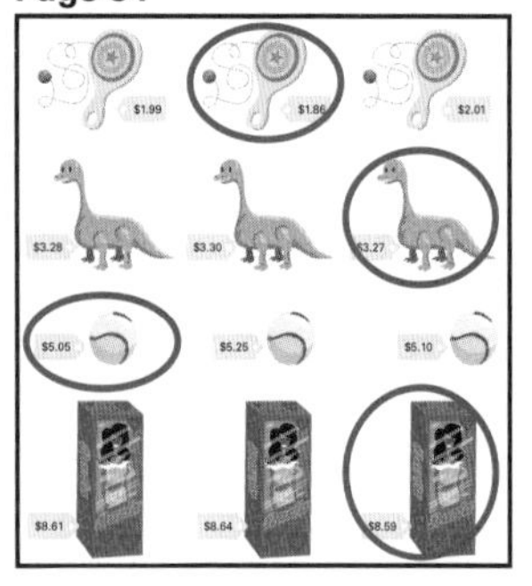

Page 32
1. 7
2. 5
3. 2
4. 3
5. 10
6. 6
7. 1
8. 4
9. 9
10. 8

Page 33
1. 18
2. 53
3. 22
4. 68
5. 1
6. 75
7. 41
8. 100

Page 34
1. 17
2. 23
3. 21
4. 19
5. 20
6. 22
7. 7.2
8. 5.9
9. 6.6
10. 6.0
11. 6.7
12. 7.4
13. 12.93
14. 13.05
15. 12.88
16. 14.22
17. 13.49
18. 12.72

Page 35
1. 89.99
2. 5.83
3. 170.37
4. 613.07
5. 29.48
6. 244.94

Page 36

Page 37

Page 38

Page 39
1. 5.60
2. 0.60
3. 4.80
4. 8.25
5. 8.75
6. 67.50
7. 25.68
8. 66.96

Page 40
1. 18.95
2. 25.30
3. 26.13
4. 29.44
5. 32.11
6. 47.20
7. 62.93
8. 67.71
9. 38.18
10. 27.78
11. 24.47
12. 18.31

Page 41
1. 1.37
2. 3.36
3. 1.84
4. 2.49
5. 1.99
6. 0.29

Page 42

Page 43

1. GUEST CHECK 000901

Shrimp scampi	$18.20
Rigatoni with broccoli	$11.99
Spaghetti and meatballs	$15.50
Food Total	$45.69
Tax	$ 8.22
Total	$53.91

Change: $46.09

2. GUEST CHECK 000902

Prime rib	$22.95
Roast chicken	$17.95
Sirloin steak	$29.95
Food Total	$70.85
Tax	$14.17
Total	$85.02

Change: $14.98

Page 43 (continued)

3. GUEST CHECK 000903

Curry stew	$13.80
Spicy rice noodles	$18.25
Spinach wonton soup	$ 6.88
Food Total	$38.93
Tax	$ 5.84
Total	$44.77

Change: $55.23

4. GUEST CHECK 000904

Tomato and feta salad	$10.35
Lamb shish kebab	$18.95
Chicken souvlaki	$15.40
Gyro	$13.24
Food Total	$57.94
Tax	$11.59
Total	$69.53

Change: $30.47

Page 44
1. 3.926 + 0.187 = 4.113
2. 216.8 + 35.91 = 252.71
3. 72.516 − 9.828 = 62.688
4. 14.5 − 5.93 = 8.57
5. 87.03 × 9 = 783.27
6. 15.84 × 1.7 = 26.928
7. 0.126 ÷ 18 = 0.007
8. 561.12 ÷ 21 = 26.72

Page 45

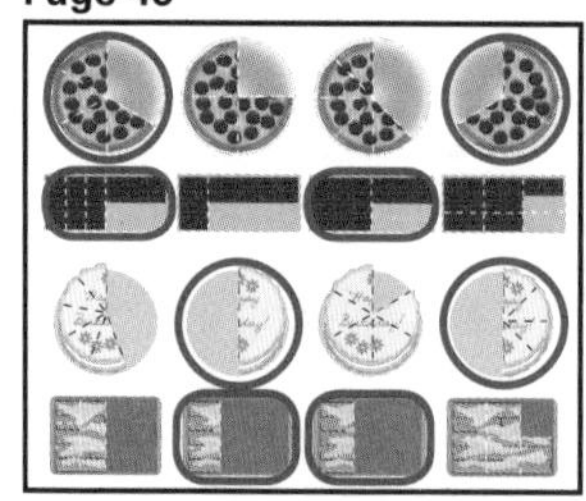

Page 46
1. $\frac{2}{2}$
2. $\frac{8}{8}$
3. $\frac{4}{4}$
4. $\frac{6}{8}$
5. $\frac{3}{3}$
6. $\frac{2}{4}$
7. $\frac{4}{6}$
8. $\frac{2}{8}$

Page 47
1. $\frac{3}{4}$
2. $\frac{2}{3}$
3. $\frac{1}{3}$
4. $\frac{3}{4}$
5. $\frac{1}{4}$
6. $\frac{1}{3}$
7. $\frac{1}{2}$
8. $\frac{2}{5}$
9. $\frac{3}{5}$

Page 48
1. $\frac{1}{2}, \frac{1}{3}, \frac{1}{6}$
2. $\frac{1}{9}, \frac{2}{9}, \frac{2}{3}$
3. $\frac{1}{3}, \frac{1}{6}, \frac{1}{2}$
4. $\frac{1}{4}, \frac{5}{12}, \frac{1}{3}$

Page 49
1. $1\frac{1}{3}$
2. $1\frac{3}{5}$
3. $2\frac{1}{2}$
4. $1\frac{1}{6}$
5. $2\frac{1}{4}$
6. $1\frac{3}{16}$
7. $1\frac{7}{8}$
8. $2\frac{3}{10}$

Page 50
1. $6\frac{1}{4}$
2. $6\frac{3}{4}$
3. $10\frac{1}{3}$
4. $8\frac{5}{6}$
5. $7\frac{1}{2}$
6. $11\frac{2}{3}$
7. $4\frac{1}{4}$
8. $10\frac{3}{8}$

Page 51
1. $\frac{3}{2}$
2. $\frac{10}{3}$
3. $\frac{7}{3}$
4. $\frac{11}{2}$
5. $\frac{7}{4}$
6. $\frac{17}{8}$
7. $\frac{14}{3}$
8. $\frac{15}{8}$

Page 52
1. $\frac{19}{8}$
2. $\frac{11}{6}$
3. $\frac{15}{4}$
4. $\frac{19}{10}$
5. $\frac{10}{3}$
6. $\frac{9}{4}$
7. $\frac{9}{2}$
8. $\frac{23}{6}$
9. $\frac{45}{8}$
10. $\frac{25}{9}$
11. $\frac{25}{8}$
12. $\frac{37}{6}$

Page 53
1. $\frac{32}{100}$, 0.32, 32%
2. $\frac{15}{100}$, 0.15, 15%
3. $\frac{53}{100}$, 0.53, 53%

Page 54
1. $\frac{4}{10}$, 0.4, 40%
2. $\frac{8}{10}$, 0.8, 80%
3. $\frac{7}{10}$, 0.7, 70%
4. $\frac{1}{10}$, 0.1, 10%

Page 55
1. 99
2. 75
3. 80
4. 60
5. 47
6. 30
7. 48
8. 2
9. 50
10. 4
11. 92
12. 20

Page 56
1. 30
2. 50
3. 15
4. 20
5. 35

Page 57

GUEST CHECK 000946

Meatball sub	$ 6.20
Bag of chips	$ 1.65
Bottle of water	$ 2.35
Food Total	$10.20
10% tip	$ 1.02
Total	$11.22

GUEST CHECK 000947

Chicken club sandwich	$ 8.55
Bowl of soup	$ 3.20
Small coffee	$ 0.85
Food Total	$12.60
15% tip	$ 1.89
Total	$14.49

GUEST CHECK 000948

Grilled ham and cheese	$ 4.15
Lemonade	$ 2.40
Cherry pie	$ 3.30
Food Total	$ 9.85
20% tip	$ 1.97
Total	$11.82

GUEST CHECK 000949

Garden salad	$10.25
Iced tea	$ 2.60
Chocolate pudding	$ 1.75
Food Total	$14.60
25% tip	$ 3.65
Total	$18.25

Page 58
1. 32 2. 18 3. 20
4. 8 5. 12 6. 50
7. 15 8. 20 9. 40
10. 510 11. 780 12. 210

Page 59
1. $\frac{1}{2}$ 2. 0.6 3. 40%
4. $\frac{3}{5}$ 5. 50% 6. 0.48

Page 60

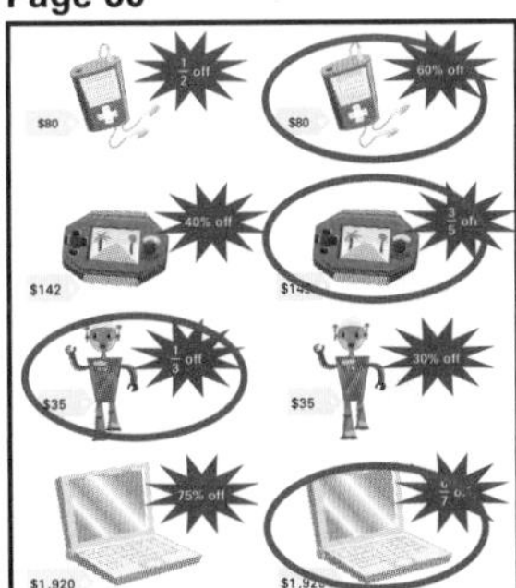

Page 61
1. Sandra 2. Alex
3. Melissa 4. Marcus's

Page 62
1. 288 2. 280 3. 247
4. 225 5. 238 6. 264
7. 270 8. 221 9. 231

Page 63
1. $\frac{3}{5}$ 2. $\frac{7}{9}$ 3. $\frac{2}{3}$
4. $\frac{1}{2}$ 5. $\frac{3}{5}$ 6. $\frac{5}{6}$
7. $1\frac{1}{7}$ 8. $1\frac{1}{2}$

Page 64
1. $1\frac{1}{4}$ 2. $1\frac{5}{12}$ 3. $\frac{5}{6}$
4. $1\frac{1}{12}$ 5. $\frac{7}{12}$ 6. $\frac{3}{4}$

Page 65

Page 66
1. $10\frac{4}{5}$ 2. 16 3. $22\frac{1}{4}$
4. $13\frac{5}{8}$ 5. $17\frac{3}{14}$ 6. $24\frac{1}{2}$
7. $19\frac{1}{6}$ 8. $29\frac{1}{9}$

Page 67
1. $\frac{1}{2}$ 2. $\frac{1}{3}$ 3. $\frac{3}{8}$
4. $\frac{2}{3}$ 5. $\frac{1}{5}$ 6. $\frac{2}{3}$

Page 68
1. $\frac{1}{2}$ 2. $\frac{1}{2}$ 3. $\frac{3}{4}$
4. $\frac{1}{12}$ 5. $\frac{3}{20}$ 6. $\frac{5}{8}$
7. $\frac{4}{15}$ 8. $\frac{11}{40}$

Page 69

Page 70
1. $3\frac{1}{2}$ 2. $8\frac{1}{8}$ 3. $1\frac{1}{4}$
4. $4\frac{1}{12}$ 5. $5\frac{7}{8}$ 6. $\frac{7}{12}$

Page 71
1. $1\frac{1}{4}$ 2. $3\frac{3}{4}$ 3. $1\frac{2}{3}$
4. $2\frac{1}{2}$ 5. $3\frac{1}{3}$ 6. $4\frac{1}{6}$

Page 72
1. 4 2. $1\frac{1}{4}$ 3. $1\frac{1}{2}$
4. $\frac{1}{2}$ 5. 3 6. 1
7. $1\frac{3}{5}$ 8. $\frac{2}{3}$ 9. 8
10. $2\frac{1}{2}$ 11. 3 12. 1
13. 6 14. 2 15. $3\frac{1}{5}$
16. $1\frac{1}{3}$

Page 73

Page 74

Part	Days	Miles	Total Miles
1	40	$2\frac{1}{2}$	100
2	35	$4\frac{1}{5}$	147
3	30	$6\frac{3}{5}$	198
4	24	$10\frac{2}{3}$	256
5	18	$15\frac{5}{6}$	285
6	12	$18\frac{3}{4}$	225
7	7	$22\frac{4}{7}$	158
8	5	$26\frac{2}{5}$	132

Page 75
1. 48 2. 18 3. 192
4. 40 5. 18 6. 3
7. 5 8. 6 9. 3
10. 9

Page 76
1. $\frac{1}{8}$ 2. $\frac{1}{10}$ 3. $\frac{1}{14}$
4. $\frac{1}{8}$ 5. $\frac{2}{9}$ 6. $\frac{1}{15}$
7. $\frac{3}{20}$ 8. $\frac{2}{7}$

Page 77

Page 78
1. 94 2. 15 3. 22
4. 13 5. 34 6. 56

Page 79
$2\frac{3}{4}$, 11

Page 80
1. $\frac{1}{2}$ 2. $\frac{1}{6}$ 3. $\frac{1}{24}$
4. $\frac{1}{8}$ 5. $\frac{1}{4}$

Page 81
1. $30\frac{1}{4}$ 2. $35\frac{3}{4}$ 3. $41\frac{1}{4}$
4. $49\frac{1}{2}$ 5. 13 6. 4

Page 82
1. 3.8 2. 2.6 3. 4.2
4. 26 5. 38 6. 23

Page 83
1. 0, 3, 2 2. 3, 0, 0
3. 4, 0, 0 4. 0, 2, 2
5. 0, 3, 0 6. 0, 1, 2

Page 84
Suggestion:

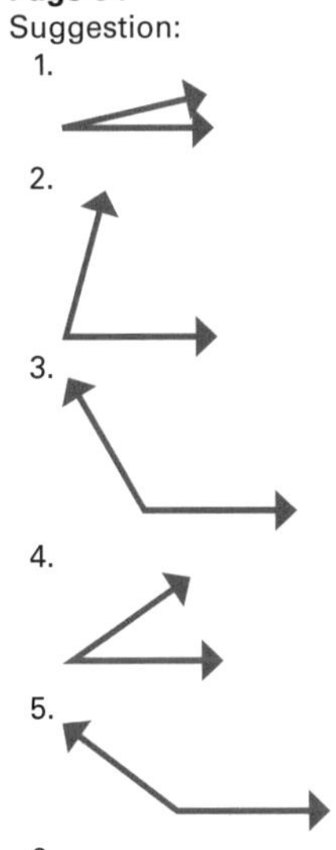

Answers

Page 85

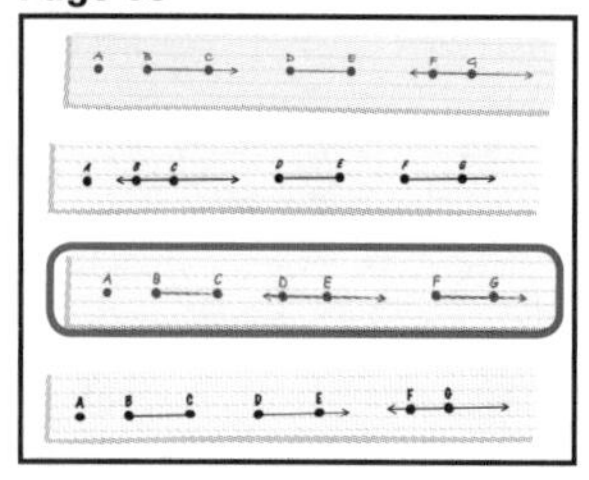

Page 86

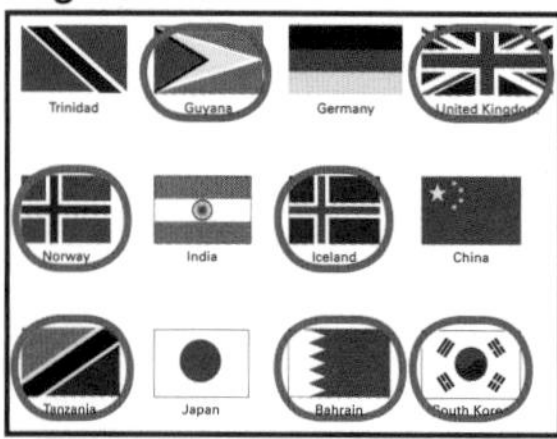

Page 87

Suggestion:

1.
2.
3.
4.
5.
6.
7.
8.
9.

Page 88

❑ nine vertices
☑ shape name: heptagon
☑ seven obtuse angles
❑ two right angles
☑ no pairs of parallel lines

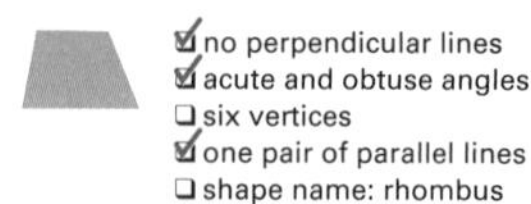

☑ no perpendicular lines
☑ acute and obtuse angles
❑ six vertices
☑ one pair of parallel lines
❑ shape name: rhombus

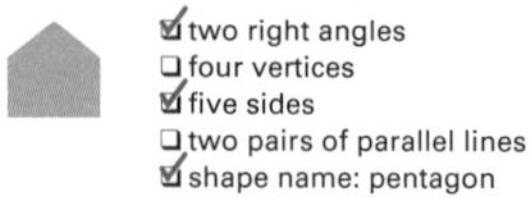

☑ two right angles
❑ four vertices
☑ five sides
❑ two pairs of parallel lines
☑ shape name: pentagon

☑ shape name: parallelogram
☑ four vertices
❑ four right angles
☑ two pairs of parallel lines
☑ two obtuse angles

Page 89

1. bedroom 4
2. bedroom 1
3. 66 feet

Page 90

1. 60
2. 41.6
3. 25.8
4. 74.8
5. 202.2

Page 91

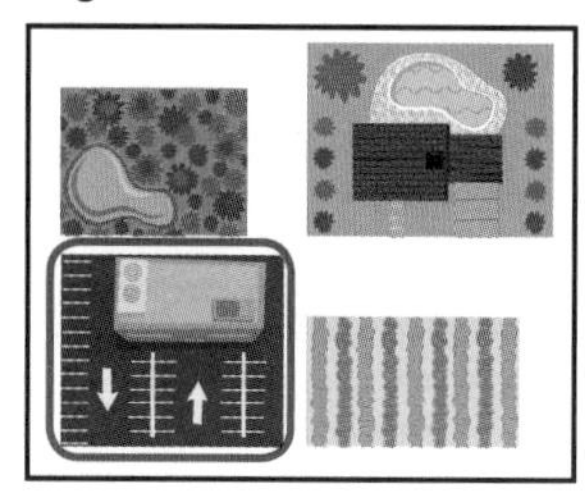

Page 92

33.81

Page 93

1. Chocolate Cherry
2. Truffle
3. Marshmallow
4. Toffee
5. Orange Crème
6. Caramel

Page 94

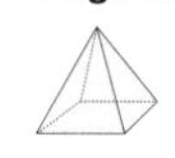

☑ shape name: square pyramid
☑ 5 vertices
❑ 6 edges
☑ 1 square face
❑ all faces that are triangles

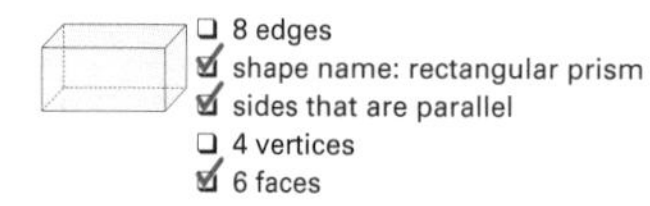

❑ 8 edges
☑ shape name: rectangular prism
☑ sides that are parallel
❑ 4 vertices
☑ 6 faces

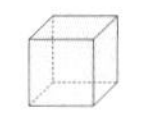

☑ shape name: cube
❑ no parallel lines
☑ 12 edges
☑ 8 vertices
☑ all square faces

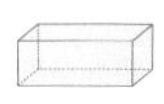

❑ 8 faces
☑ all faces that are rectangles
❑ shape name: cube
☑ 12 edges
☑ 8 vertices

Page 95

1. 14   2. 10
3. 14   4. 22

Page 96

1. 1,800   2. $1{,}827\frac{1}{2}$
3. 1,518   4. 861

Page 97

1. parakeet   2. ferret
3. guinea pig   4. hamster

Page 98

2

Page 99

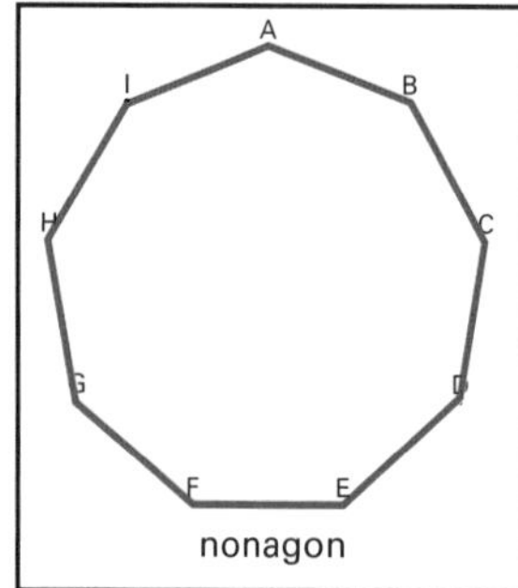

Page 99 (continued)

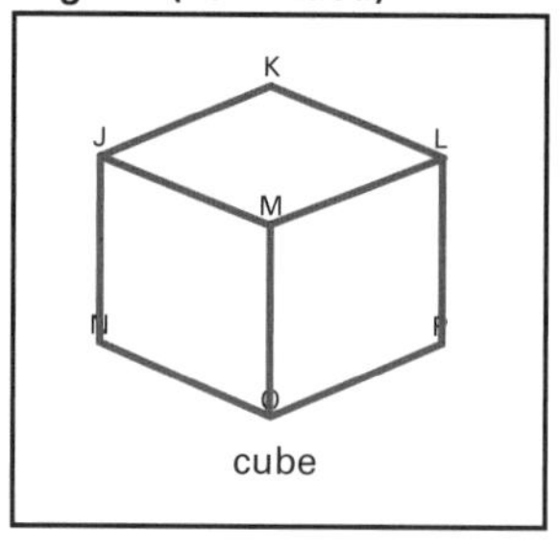

Page 100

1. 25
2. 37.5
3. 318.75

Page 101

1. 6   2. 2
3. trumpet   4. drums, bass
5. 7   6. 4
7. 9   8. 39

Pages 102–103

1. 6   2. 5
3. 4   4. 5

Page 104

Have someone check your answers.

Page 105

1. 25,000
2. 40,000
3. 20,000
4. Zoo, Children's Museum, Art Museum
5. City History Museum
6. 105,000
7. 70,000
8. 40,000

Page 106

1. 4   2. 2
3. game 1   4. game 4
5. game 5   6. 7
7. 3   8. 2

Page 107

1. 700,000
2. 1,000,000
3. Sonja Gonzales
4. Brad Thompson
5. 1,500,000
6. 500,000
7. 300,000
8. Sonja Gonzales

Page 108

Have someone check your answers.

Page 109

1. 50%   2. 25%
3. basketball   4. track
5. soccer   6. track
7. unlikely   8. impossible

Page 110

1. Ages 18–34
2. Ages 35–49
3. Ages 9–17

Page 111

1. 115
2. 75
3. 110
4. 50
5. 35
6. 25
7. 90
8. Superhero Supreme, Meatball Mayhem
9. Cool Club, Bacon Bonanza

Page 112

Have someone check your answers.

Page 113

1. Jacob
2. Ethan
3. Jennifer, Jacob
4. Jacob, Ethan
5. Ethan
6. Elizabeth
7. 1990
8. 2000

Pages 114–115

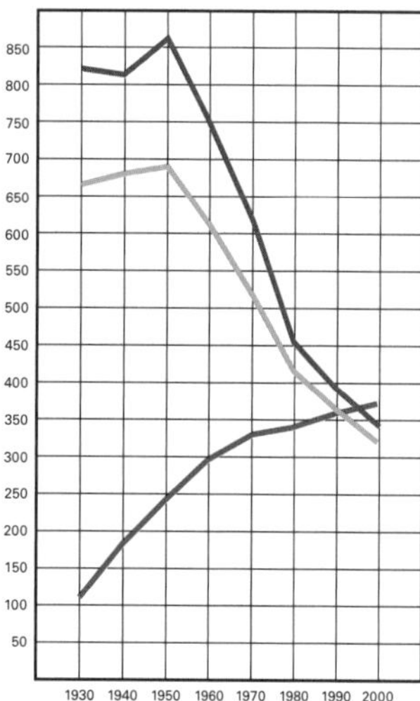

Page 116

1. 350   2. 200
3. Mitch   4. Mitch
5. April   6. unlikely
7. likely   8. 285

Page 117

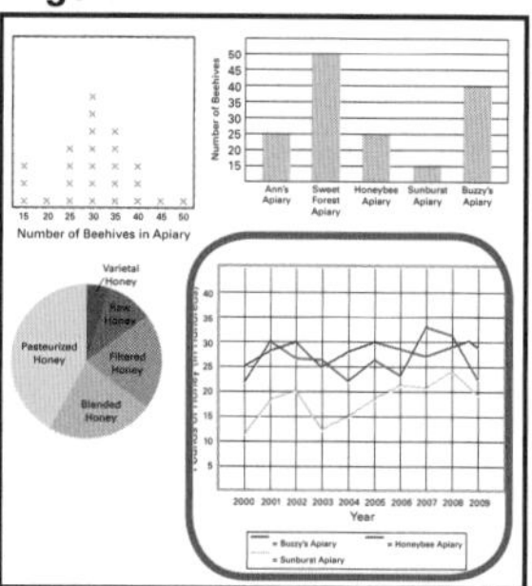

Page 118

Have someone check your answers.